Materials Horizons: From Nature to Nanomaterials

Series Editor

Vijay Kumar Thakur, School of Aerospace, Transport and Manufacturing, Cranfield University, Cranfield, UK

Materials are an indispensable part of human civilization since the inception of life on earth. With the passage of time, innumerable new materials have been explored as well as developed and the search for new innovative materials continues briskly. Keeping in mind the immense perspectives of various classes of materials, this series aims at providing a comprehensive collection of works across the breadth of materials research at cutting-edge interface of materials science with physics, chemistry, biology and engineering.

This series covers a galaxy of materials ranging from natural materials to nanomaterials. Some of the topics include but not limited to: biological materials, biomimetic materials, ceramics, composites, coatings, functional materials, glasses, inorganic materials, inorganic-organic hybrids, metals, membranes, magnetic materials, manufacturing of materials, nanomaterials, organic materials and pigments to name a few. The series provides most timely and comprehensive information on advanced synthesis, processing, characterization, manufacturing and applications in a broad range of interdisciplinary fields in science, engineering and technology.

This series accepts both authored and edited works, including textbooks, monographs, reference works, and professional books. The books in this series will provide a deep insight into the state-of-art of Materials Horizons and serve students, academic, government and industrial scientists involved in all aspects of materials research.

More information about this series at http://www.springer.com/series/16122

Harvinder Singh
Department of Civil Engineering
Guru Nanak Dev Engineering College
Ludhiana, Punjab, India

ISSN 2524-5384 ISSN 2524-5392 (electronic)
Materials Horizons: From Nature to Nanomaterials
ISBN 978-981-16-3210-5 ISBN 978-981-16-3211-2 (eBook)
https://doi.org/10.1007/978-981-16-3211-2

This Springer imprint is published by the registered company Springer Nature Singapore Pte Ltd.
The registered company address is: 152 Beach Road, #21-01/04 Gateway East, Singapore 189721, Singapore

Harvinder Singh

Structural Materials

Behavior, Testing and Evaluation

Springer

Preface

'Materials, of themselves, affect us little; it is the way we use them which influences our lives' (*Epictetus, AD 50–120*); these are the natural gift to the mankind to live their life comfortably during their stay on the planet Earth. However, the last few decades saw a systematic growth of organized research efforts by great minds to identify, modify and synthesize different materials. These efforts led to the discovery of wonderful new materials and expanded the list containing only the natural to numerous man-made materials along with the formulation of material models and analysis methods to capture their response. The development of the new materials especially in the last 5–6 decades has formed the backbone on many pathbreaking inventions that helped the human civilization to scale new heights and live comfortably in comparison with the ancient time when they had only a few natural materials at their disposal. This monograph attempts to describe different types of materials, their classifications and insights about the fundamental mechanisms that give them distinct characteristics. It also presents the material response that they would exhibit under the routine loading conditions, generally found in civil engineering projects. The internal stress transfer mechanisms and the resultant properties that the materials exhibit are also described.

This monograph is a modest attempt to explore the world of materials as it applies to their applications as structural materials in the context of civil engineering. It attempts to summarize the vast amount of the scientific understanding and other research information related to the behavior and the response of different materials, such as the metals, nonmetals and composites at one place. I hope it will prove useful to the budding engineers to develop a taste to explore them into the deeper waters. At the teaching level, the book is intended as the text for the second year of the undergraduate civil engineering program. It will also be helpful to the academicians and the researchers who want to refresh and understand the response of different materials and how these can be manipulated to make them better or make them behave as per the project needs. For the convenience of the readers, the notation/symbols are given and explained at the point of their use in the text.

Before concluding the Preface, I would like to express my sincere appreciation and thanks to all those who are working continuously across the globe to explore different materials for the betterment of the humanity. The author also acknowledges his debt to various agencies, particularly the ACI, ASTM, BIS, CNR-DT, ICE, RELIM and different sources available on the Internet, for their published material and various images/photographs, respectively, to which references are made in the monograph for the benefit of the readers. Every effort has been made to identify copyright holders and obtain their permission for the use of copyright material. Notification of any additions or corrections that should be incorporated in future reprints or editions of this book would be greatly appreciated.At the end, the author would like to express his heartiest appreciation for all those at Guru Nanak Dev Engineering College, Ludhiana, who rendered help and support in this work and Mr Kandala Ramakrishna, Sr Technologist, Rolling Tech Group, TATA Steel and Mr. Sahil Sahil Aggarwal, Sr Technologist – Application Support Group, TATA Steel for providing a set of valuable information for the last chapter of the book.

Ludhiana, India Harvinder Singh

Contents

About the Author

Dr. Harvinder Singh is currently Professor at Guru Nanak Dev Engineering College, Ludhiana. He is engaged in teaching, research, and consultancy in structural engineering. He has authored numerous technical publications, one laboratory manual, and one monograph on 'Reinforced Concrete Steel Fiber Concrete' (published by Springer) and has edited five books. He is UGC Research Awardee (2014–2016) and State Technical Auditor (STA), Chartered Engineer and Structural consultant involved in planning and design of various infrastructure projects and is a member of several professional bodies. He is an editorial board member and reviewer for several international journals published by ACI, ICE, fib and ASCE and completed many sponsored R&D projects. His current research interests are mathematical modelling and limit analysis of engineering structures, largely directed towards solving practical problems and improving the related design practices.

Chapter 1
Materials

The material is a physical substance from which different things are made for various uses of the mankind in their day-to-day activities. The very common natural material the human has been used since ancient times is wood. It is available in abundance in the forests and is used by them for making shelter, as a fuel for cooking food, as storage bins, as weapons to protect themselves, and later on, it was used to transport other things of their use from one place to another in the form of wooden carts and also as agricultural tools; the list is endless. The common characteristics that promoted the use of wood among the people are that it is a solid, and it can be given any desired shape and size depending upon the need and its weight is quite less, which was not possible in other materials available at their disposal at that time (e.g., stone, asphalt/bitumen, water etc.). The people then must have examined these characteristics of the wood before adopting it in their life and use it for various activities. In many other applications, however, stones found its use only next to the wood where they need something stronger and durable than the wooden items. Nonetheless, its dressing was not that easy as the wood was permitting them to make it suitable for different uses; still the stone was used in many applications, such as weapons, agricultural tools, houses, pools to store water, etc. Similarly, the bitumen—a naturally occurring organic byproduct of decomposed plants usually available in the form of a black, oily, viscous liquid—was another material that found its use for various applications, such as sealant, adhesive, for igniting the fire, as a building mortar and as a decorative pigment on pots, buildings or human skin. It is still in use, albeit for different purposes, mainly among them is in road and parking floor construction. Another important natural material discovered and used by humans was oil shale—an organic-rich fine-grained sedimentary rock. It had been used as a fuel to ignite fire since prehistoric times, as it burns easily without any processing. This way, the humans go on exploring many other materials in their surroundings that meet their requirements and the result is that we have now an endless list of materials to select from.

H. Singh, *Structural Materials*, Materials Horizons: From Nature to Nanomaterials,
https://doi.org/10.1007/978-981-16-3211-2_1

The above paragraph shows that all materials possess a set of common characteristics that can be easily used to classify them and help users to select appropriate material for meeting their requirements. Being a part of matter, the materials can exist in liquid, solid and gaseous form. Water finds its use as material in liquid form in addition to numerous other materials, such as oil and paints. You can easily make a list of such materials possessing liquidity (try this). Stone, wood, soil, asphalt, metals (such as, gold, silver, iron, aluminum) find their use as solids; all these exist in the nature and are used by the humans just by dressing them to a suitable form, shape and size to meet their needs. Air is a form of the matter that finds its use in gaseous form and again exists freely in the nature. All those materials that are available in the nature freely and serve one or other purposes of use in meeting the needs of users, usually with little or no dressing are called as natural materials. Figure 1.1 depicts the use of the matter in its various forms. These can exist as pure or impure, living or nonliving matter. Matter becomes material if it is of any use to the human in meeting their one or other requirements of day-to-day life and/or future activities.

Materials can be compared and categorized by means of different quantitative measures of their response exhibited by them under various physical conditions, viz-a-viz forces, strains, temperature variations, electric current/potential, magnetic field, etc. Alternatively, it can be done on the basis of the place of their geological origin and use; or physical and chemical properties; or on the basis of biological

Fig. 1.1 Natural materials used by humans for meeting their different needs: **a** wood, **b** asphalt, **c** stone, **d** water stored in underground tanks constructed using stones after their dressing and joined together by means of any binder, such as limestone powder/asphalt, etc. *Source* Author

function. The changes in the material response or changes in the material itself with time are also some of the measures used to define and classify the matter. Basically, everything that exists in the nature is relative. The changes that materials undergo during their lifespan can be observed or these remain non-observable; it all depends upon the time-frame selected to observe them. Some materials undergo changes very rapidly and observers can take notice of these in the same time interval, and sometimes, these did not come to the notice and becomes measurable only over a very long period of time, even beyond the normal life span of the observer. All materials being used for the construction projects, mostly comprise of nonliving matter, means these did not display the key characteristics of the life, such as the ability to grow, breath, reproduce, take in and use energy and excrete waste, but exhibit many other characteristics worthy to use them for various purposes, like load bearing capacity, insulation (of space from heat and sound), conductance (of heat and electricity), permeability (against movement of water and air through them), the color and texture, ability to retain their shape and sizes, the resistance to deformations, the capacity to absorb energy, durability and many more. All these parameters remain more or less constant over a normal life span of the buildings and other infrastructure projects like bridges, roads, dams, etc. and provide a quantifiable list of attributes that can be taken as a basis to select a suitable material depending upon the type of the project. It is the most important task and/or duty of engineers to choose the right material from an endless list of materials for any job/ construction project that they pursue or undertake. Not choosing the right material often leads to unintended consequences, like frequent repair and/or lose of the load carrying capacity of members made up from such materials, or not getting a performance that they were expecting, etc. It is therefore very important for them to know as many materials as possible, their behavior under different type of loading conditions and/or environmental conditions and other weakness to enable them to select the most appropriate material for a given set of field/design constraints.

1.1 Material Classification

Materials are broadly classified as: (1) Natural and (2) man-made. All those materials that exist in the nature and can be used as such for meeting some purpose, albeit with some dressing or changes in their shape or size to make them fit for use are classified as natural materials, e.g., air, soil, stones, water and wood; a few of them are shown in Fig. 1.1. On the other hand, if some material needs processing and some modifications in their features (physical or chemical or both) to make them fit for use are called as man-made materials. Normally, these are manufactured in the factories by processing the raw materials, mainly available in the form of ores, etc., to convert them to their more useful physical or chemical form. The typical examples are metals (derived from their ores), bricks (made from the soil), plywood (manufactured from the natural wood), cement (produced from the limestone), etc.; these are depicted in Fig. 1.2.

Fig. 1.2 Man-made materials produced from various natural materials by modifying/blending/purifying them: **a** metals, **b** bricks, **c** plywood, and **d** cement *Source* Author

Based upon the characteristics and their usefulness for different engineering applications, the man-made materials are further divided into two types: (1) Engineering materials and (2) nonengineering materials, also called as functional materials.

The materials which are mainly used to meet some engineering purpose in a project are called as engineering materials; the purpose of their use can be their ability to transfer the load, to retain the pre-assigned shape and size, insulation and conductance characteristics, durability, etc. All these capabilities exist in them because of their mechanical, physical, chemical and manufacturing properties. Whereas, if the materials find their use for a purpose other than the engineering or it can be just to protect the engineering properties of some other (load bearing) materials being used in a project are called as nonengineering materials. The purpose in this case can be to get a desired texture, color, smoothness, reflection, perviousness, etc. Their common examples are paints, asphalt, waterproofing materials, acoustical materials, etc. In any construction projects, we need both engineering materials as well as nonengineering materials to get the desired results and also to optimize the speed and cost of a construction project. But it is always cheaper and sustainable to use the natural materials to the maximum possible extent and it should be preferred and used in place of the man-made materials wherever their use is found feasible, technically or economically or both.

Depending upon the distinct characteristics exhibited by these materials, these can further be grouped into three major types: (1) Metals, (2) non-metals and (3) composites.

Metals compose about three-fourth of all known elements, but only few of them find their use in the pure form in routine engineering applications. They normally exist in the solid states at the normal temperature; some metals, however, do exist in the liquid state (e.g., Mercury) at the room temperature, Metals are usually lustrous, ductile, malleable, good conductor (of heat and electricity), strong and ductile (see, Fig. 1.2a). Some examples of metals are aluminum, copper, gold, iron, lead, nickel, silver and zinc. Mostly, the metals are blended with other elements to improve their existing natural traits and make them fit for some other engineering applications; the metals in their modified form are known as metallic materials (e.g., alloys). The availability of a large number of non-localized electrons, i.e., the electrons which are not bound to particular atoms, both in the metals and metallic materials, is responsible for the various properties exhibited by them; this allows them to readily give up the electrons to form metallic bonds, which impart them the capability to bear the load and deform along with the ability to conduct the heat and electricity and also make them more lustrous relative to other materials in the group.

The metallic materials are further of two types; (1) Ferrous, (2) Non-ferrous. A group of the metallic materials which mainly composed of iron (Fe) as a constituent unit of the material is called as ferrous metals (e.g., cast iron; steel—mild steel, carbon steel, stainless steel, and wrought iron); otherwise, these are known as non-ferrous metals (e.g., aluminum, copper, gold, lead, nickel, silver, etc.). Both of these groups are good conductors of the heat and the electricity; however, only the former group possesses magnetic properties. Another feature of these two groups is that they easily tend to convert back to their original form, i.e., ore from which these have been obtained in the pure form through the process of oxidation. This is the biggest disadvantage of using the metals in their pure form. For instance, iron if used in the form of reinforcing steel bars in (porous) concrete gets easily converted to the iron oxide (a form of the iron ore) once it gets in touch with the air and moisture; this process in common parlance is called as rusting/corrosion. It is a form of the metal loss, which sometimes results in the degradation of desired load bearing capacity of members if it is coming mainly from that embedded metal, e.g., the steel bars embedded in concrete impart the member a capacity to resist tension. Engineers therefore have to be very conscious while selecting materials that their improper/inappropriate selection should not lead to the strength degradation or lose of other desirable properties of objects or members under considerations.

Unlike the metals, the non-metals are non-crystalline in nature. Normally, these are also bad conductors of the heat and the electricity, which makes them most suitable materials for the insulation purpose. At the room temperature, these can exist in both solid and gaseous forms. Figure 1.3 depicts different non-metals that are used commonly in various civil engineering applications. Some of the examples of non-metals are asbestos, cotton, ceramic, leather, nylon, polymers, plastics, rubber, etc.

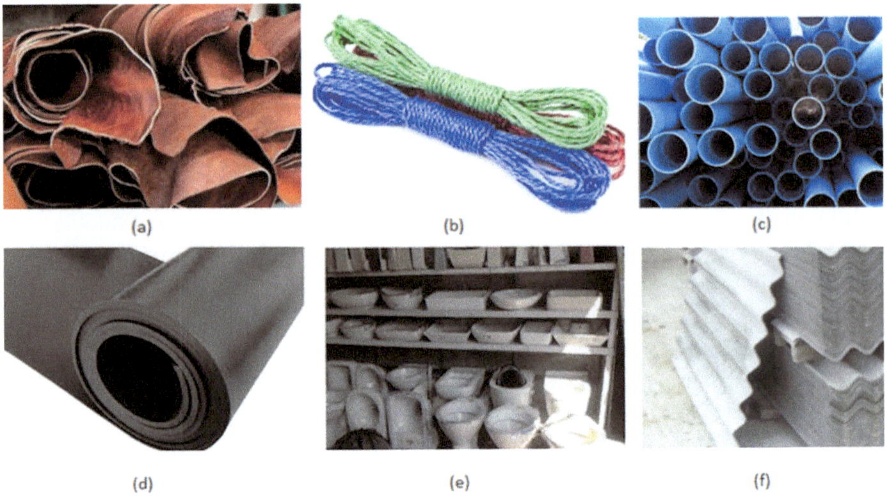

Fig. 1.3 Non-metals produced from other materials by modifying/blending them: **a** leather, **b** nylon, **c** polymers (in the form of pipes), **d** rubber sheet, **e** ceramics (in the form of sanitary fittings), **f** asbestos (in the form roofing sheets). *Source* Author

Composites are the third major group of the engineering materials. These are produced by blending together two or more materials (may be from a list of the natural or the man-made group) at the time of their manufacturing/production with a sole purpose to improve one or more engineering properties of the composite product/material so formed, by incorporating the best characteristics of each constituent materials. The purpose of the blending may be to reduce the unit weight of the final required material, and/or to improve their structural properties, such as strength, toughness, stiffness, and/or to make them fire and corrosion resistant, and/or to modify their conductance properties, and/or to enhance their durability characteristics, etc. Basically, the composites are designers' materials, especially, tailor-made to meet the needs of a project. The composite provides engineers a means of achieving high strength at the lower weight because they usually exploit very strong fibers embedded in the light matrices. Engineers depending upon the need of the project explore and zero-in on a set of the design constraints and try to plan and design the project in hand using the available materials nearby may be from a group of natural or man-made materials. If these materials, in isolation, fail to serve the intended purpose, then some other material(s) from the list of composites can be selected. Sometimes, the reduction of project cost may be another factor that weighs in favor of the composites. A few typical examples of the engineering composites are plywood; fiberglass; reinforced materials; metal-matrix, ceramic-matrix, and polymer-matrix composites; etc.

For instance, the wood is a natural material that consists of strong and flexible cellulose fibers and composite produced from it—Plywood—consists of a number of thin sheets of wood, normally bonded together by polymer in between them to enable them to act as a single unit with improved properties. It is a form of sandwiching in which different layers/sheets, often very thin, of woods are placed one over the other and glued together by means of some adhesive to achieve the desired material properties/performance level (see, Fig. 1.2c). Similarly, the glass fibers if considered alone as single or strands are strong (but, are brittle) but if these are embedded in the polymeric materials, such as epoxy or polyester, which are normally ductile in nature (but, is weak in isolation) results in a composite—carbon fiber-reinforced polymeric material (CFRP)—having best of the two constituting materials properties, having more strength and high ductility and also, possessing very high value of the strength-to-weight ratio. Figure 1.4 depicts a typical view of the glass fibers and the composite sheet produced from such fibers. This material finds enormous applications in the aerospace industry and the high-tech sporting equipment (e.g., bicycles, tennis rackets, golf clubs, skis/snowboards, etc.) where the low weight of a member and the highest possible strength is the major requirement.

The reinforced materials are also a form of the composites, usually consisting of two or more materials—one set of the materials reinforcing the weaknesses of some other material. It can be produced by providing a stronger material called as 're-inforcement' externally on the face (s) of the member made up of some other material (less strong) or by embedding the reinforcing material (normally, stronger materials) into a weaker material at the time of the production. This type of technique is in practice since ancient times. The use of natural fibers as a reinforcing medium in the mixture of the lime and sand is the earliest known technique; the fibers were primarily added to increase the stability of the product/material or increase its load carrying capacity. Figure 1.5 shows a process of reinforcing the plain concrete, which in itself is a brittle material, but starts exhibiting ductile response once it is reinforced suitably using steel fibers and/or steel bars in the one and two orthogonal directions. Their use dates back to the Roman period (300 BC—476 AD), wherein the ancient concrete was found to contain natural fibers. Similarly, straw-reinforced mud bricks were found at a number of ancient sites in the Middle-East dated back to around 10,000 years ago. The indigenous inhabitants in the USA were using the sun dried adobe bricks, believed to be made using a

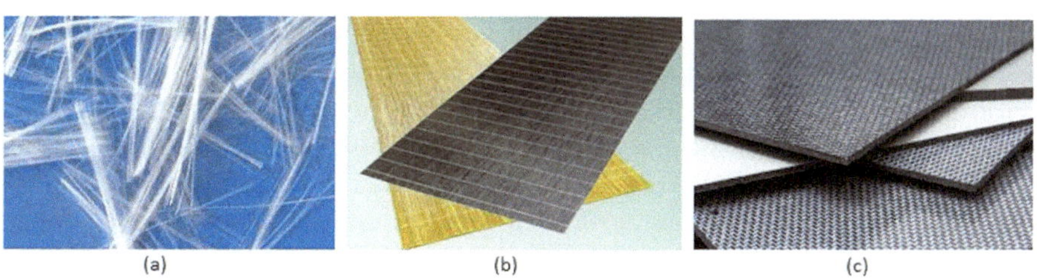

Fig. 1.4 Composites produced by reinforcing one material with another to get the best of the two in the form of enhanced mechanical properties: **a** glass fiber strands, **b** and **c** carbon fiber-reinforced polymeric (CFRP) sheets. *Source* Author

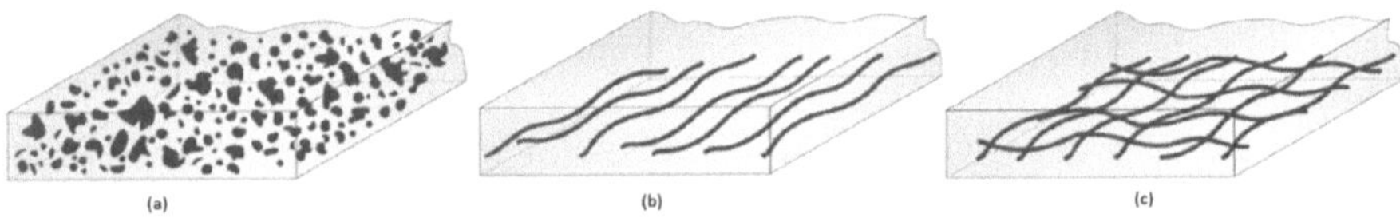

Fig. 1.5 Composites produced by reinforcing one material (usually, brittle in nature) with another one (always strong enough in tension to provide an effective bridging action) to get the better of the two in the form of enhanced post-cracking mechanical characteristics: **a** plain concrete, **b** concrete reinforced with steel bars in one direction, **c** concrete reinforced with steel bars in the two directions. *Source* Author

mixture of sand, clay and the straw. The same process is now prevalent throughout the globe, albeit in its modified form; the mix of lime and sand nowadays is changed to cement concrete and the natural fibers have been replaced with the stronger steel reinforcing bars or short steel fibers.

1.1.1 Structural and Nonstructural Materials

The materials whether natural or man-made if are used to produce/make objects with a sole purpose to transfer the load from the point of its application to some other point of the object are called as structural materials. This means that these materials must be strong enough to bear the applied load and permit its safe transfer through themselves without any failure or showing any other signs of distress. The other equally important requirement may be that it should not unnecessarily deform or undergo changes in its shape and/or size while performing such a function. These two functions are primarily defined using two basic material parameters called as its strength and stiffness. The former is the capacity of the material to bear the load just before its failure occurs while the latter parameter denotes its deformation capacity under a load, i.e., how it would deform under the action of some load. Normally, most of the materials go on deforming (may be its shortening or elongation) as the load acting over it is increased. Nonetheless, it happens either elastically followed by a sudden failure (such type of materials is termed as brittle) or it occurs elastically up to some value of the applied load, followed by a continuous increase in the deformations till its failure; this type of the material response is called as ductile behavior. Figure 1.6 depicts a typical material response in the form of a load–deformation curve.

Generally, most of the natural materials, such as stone and rocks exhibit a brittle behavior (see, Fig. 1.6a), whereas some of them undergo a partly ductile response, e.g., wood, before the failure occurs. However, through a proper blending and design, the response of materials can be easily transformed from a brittle to the ductile one (see, Fig. 1.6b). For instance, the reinforced concrete behaves as ductile materials if it is properly designed to behave like such materials. However, in case of inadequate and/or improper design/detailing, the same-reinforced concrete

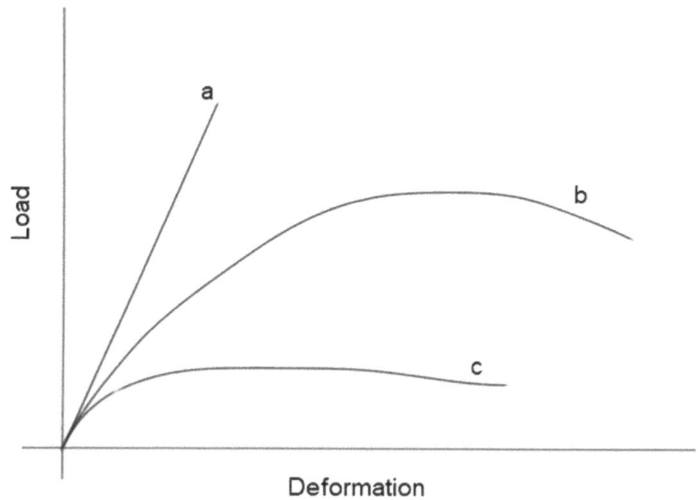

Fig. 1.6 Typical load–deformation response of materials: **a** brittle—deforms elastically followed by a sudden failure, **b** ductile—significant elastic range (with enough initial stiffness) followed by a large plastic deformations till its sudden failure, **c** plastic—very short elastic range, followed by large deformations till its failure. *Source* Author

member may result in the failure through a brittle mode. The composite material is the only hope to get rid of such type of sudden failures but is possible only if the members using such materials are properly designed to behave like the ductile materials. In these days, composite materials find enormous applications in various construction activities/projects. Reinforced concrete is one such material which is in use for the last 200 years (albeit, in its present form); although the plain concrete is being used for different activities for the thousands of years! The oldest concrete discovered was on the floor of a hut found in Israel; historically, it dated back to about 7000 BC. The concrete found in the floor of the hut is believed to be made by burning a limestone to produce the quicklime, which was somehow mixed with the water and stone particles present there and that mix finally took the shape of a solid mass after its hardening. Since then, the concrete, albeit in its modified forms, has been used for constructing many amazing things throughout its history, including architecture, infrastructure and many more.

1.2 Types of Material

Materials are that part of the matter that can be used for meeting some purpose/ objectives. The matter exists in the three different forms—solids, liquids and gasses. Although its fourth form—plasma also exists, but it has not been used as frequently as the other three states were. And, among the three states of the matter, materials, mostly in its solid form, find more applications than the other two forms. The definite shape that the solids provide is one of the reasons for its widespread applications in various engineering projects. The other two forms, namely the liquids and the gases, take the shape of its container in which these are stored. This key characteristic of the liquids and the gasses limit its use for a typical set of the applications where such features are of some help in meeting our purpose. But, in most of the civil engineering applications, the materials are either used to transfer forces/stresses while retaining their pre-assigned shape/size or just as a filler material to meet some purpose other than the force transfer; all the three forms of the matter (but, mostly solids and liquids) in one way or other help engineers to meet this objective.

The materials retain their shape as they are capable to transfer the stress from one point (of its application) to some other points in them safely. In case the materials fail to do so, they will flow like a liquid or move like a gas. In such a case, it will not be possible for them to maintain their pre-assigned shape/size, rather they will take up the shape of the container used to store them; or in the absence of any container, and they tend to move in the direction of the applied force. In doing so, either one or group of the following mechanisms gets activated in the materials that help them keep their pre-assigned geometrical dimensions of the object, i.e., shape and size:

1. Bonding (physical or chemical)
2. Friction (surface or internal)
3. Mechanical interlocking
4. Surface contact
5. Dowel action

Unlike the liquids and gasses, the molecules in the solids are packed very close to each other because of the strong intermolecular forces operating between them. It results in a very strong bond between the constituent particles of solids; as a result, they cannot move as freely as they normally do in case of liquids and gasses. This *bonding* is the result of the physical or chemical forces of attraction; it all depends upon the chemical composition of solids. It may manifest in the form of either adhesion (a force of attraction between the two particles of different natures) or cohesion (a force of attraction between the two particles of same nature). For instance, stones are weaker than the metals. Stones contain a variety of minerals with different chemical composition so accordingly, it transfers the stresses internally by means of a relatively weaker force—adhesion and cohesion; whereas in case of the metals (produced by purifying the ores), these are able to carry much higher force and exhibit a large deformation capacity than the stones, etc. because of the existence of a much stronger force of attraction, acting at an atomic level, and manifesting in the form of the metallic bonding, which is much stronger than the covalent/ionic bonding responsible for the cohesion. The resultant microstructure of solids (metals) plays an important role in dictating the response that these would exhibit under a set of externally applied loading conditions.

The solids are able to sustain the applied loads/stresses as long as these external actions do not exceed the internal forces of attraction existing within the constituent units of the material microstructure. However, the moment it occurred, the solids disintegrate into its constituent units; it can happen naturally (by means of weathering action) or by means of various man-made actions, such as different chemical or physical changes done intentionally only for that purpose. Once that happened, the solids (may be a big rock) divide themselves to a set of smaller units (pebbles, stones, etc.). In this stage, the packed pebbles/stones can retain their position/overall form by means of the *friction* and *mechanical interlock* that exists between the different particles. These two mechanisms once activated, prevent them to move apart under the action of externally applied forces; the particles normally tend to slide along an inclined plain if the angle of the slope is kept more than the safe angle (called as an angle of repose). But the force of friction and the mechanical interlock activated in the mass will not allow them to do so and they remain stable by forming a slope up to a certain angle with the horizontal. The only things the force can do at this stage is to reduce the spacing between the particles and help them to pack more closely, thereby increasing the friction and mechanical interlocking among them further. It also leads to enhanced surface-to-surface *contact* between the adjacent smaller units. So, the solids still remain in the position to transfer the vertically applied load, albeit through a modified stress transfer mechanisms—surface contact—even after the loss of their original form (may be rock, now in the form of small pebbles/stones) and degradation of other stress transfer mechanisms. The presence of some reinforcing bars across the sliding surface helps to increase or retain the load carrying capacity by preventing the free movement of the sliding block; this mechanism is called *dowel action*.

Depending upon how the constituent particles/units of solids are responding, there are five types of materials: (1) Inert materials, (2) cementitious materials, (3) piezoelectric materials, (4) polymeric materials and (5) biomaterials. (Please check spell of item 2; it is cementitious materials. It is written as ``cmaterials')

If there exists little or negligible adhesion or cohesion among different particles of materials (solids) and they transfer the externally applied load (even their own weight) solely by means of the surface contact and/or the friction and/or mechanical interlocking, such type of the materials are known as inert materials. For example, fine aggregates (sand) and coarse aggregates (*Bajri*)—formed by weathering of the stones (which in turn, are also formed by weathering of rocks/mountains)—are inert materials (see, Fig. 1.7).

On the other hand, if the adhesion and/or cohesion are the sole mode of their stress transfer capability, these are called as cementitious materials. For example, the cements, fly ash, lime, etc. are used as binder in different composite materials, such as concrete due to their binding characteristics which help them to develop adequate bonding with the adjoining particles in the matrix/lattice to give a solid form to the material (i.e., concrete) produced from their use, after the hardening process. Figure 1.8 shows a typical photograph of various commonly used materials in various construction activities that possesses binding properties. Clay is another example of the natural materials (e.g., soils) which form a very hard

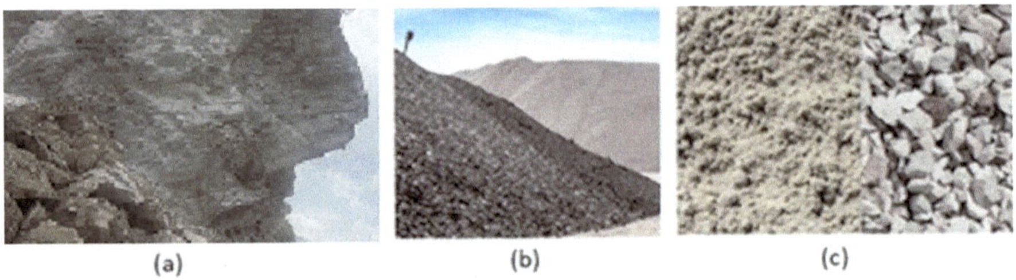

Fig. 1.7 Inert materials: **a** stones transferring their weight by means of the surface contact and the aggregate interlock, **b** a heap of aggregates forming a natural stable slope depending upon their type, **c** a close up of fine and coarse aggregates *Source* Author

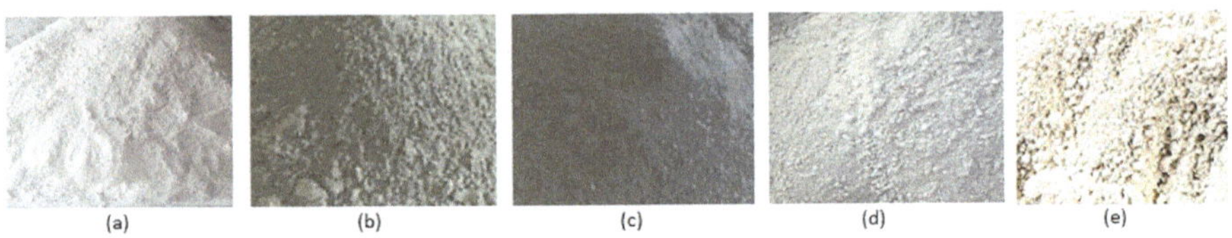

Fig. 1.8 Cementitious materials (in powder form): **a** lime, **b** fly ash, **c** silica fume, **d** cement, and **e** clay *Source* Author

substance in its dry form mainly because of the electrostatic charges that develops over the surface of the minerals it is composed of and helped these particles develop cohesion forces large enough to hold them together in the form of a solid mass.

There is another group of materials that possess unique *pozzolanic* properties because of the presence of siliceous or siliceous and aluminous content in them, which, in themselves, possess little or no cementitious value, but they will do so in the finely divided form. These materials, in the presence of water, react chemically with calcium hydroxide finding its ways into the mixture from some other sources to form the compounds possessing cementitious properties. This process happens at ordinary temperature and helps the finely powdered material to get a solid shape after the completion of the hardening process.

Piezoelectric materials are special type of materials that have a unique ability to generate an internal electrical charge when these are mechanically stressed. Some solids because of their unique internal microstructure exhibit electric-dipole moments in the lattice that produces an electric charge when some stress field tends to realign it. The charge however is produced as long as some stress field is acting over the substance, so as soon as it is removed, there is no charge produced in the material. Amazingly, the piezoelectric process is reversible in nature, so if an electric current is applied to these materials, it will cause them to change their shape; this process is schematically depicted in Fig. 1.9. Because of this unique capability, the piezoelectric materials find enormous applications in the production of different types of sensors. The sensors are used in many types of measurement equipments where the purpose is to capture and quantify the effect of some physical

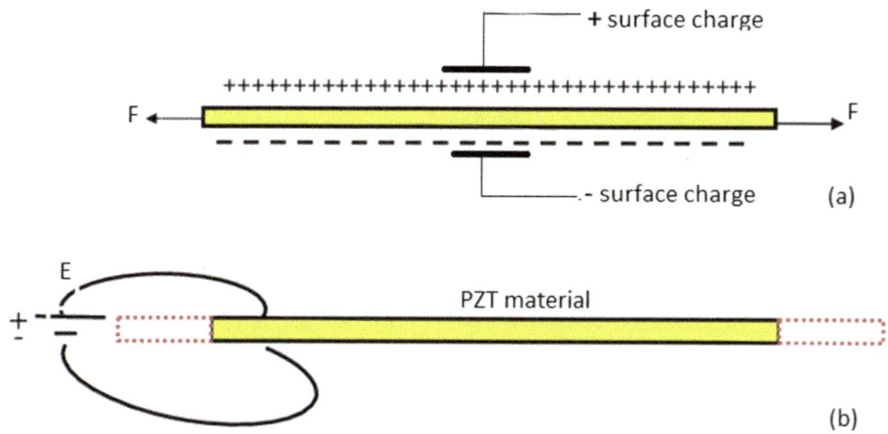

Fig. 1.9 Piezoelectric effect: **a** mechanical force, **F** generates a surface charge (+, −) on the opposite faces of a piezoelectric material (PZT) patch, **b** electric field, **E** produces a change in length of the PZT patch. *Source* Author

or electrical disturbances; the inherent nature of these materials helps them to transform the physical disturbance/stress field into a set of the electrical signals, and vice versa that can further be processed by means of some hardware and software into any useful data. For instance, the deformations undergone by solids under the action of some loads can easily be captured using sensors made up of piezoelectric materials. There are many materials which have been found to possess the piezo-electricity effect; some of these are quartz, rochelle salt, topaz, zinc oxide berlinite, langasite, lithium niobate, lithium tantalate, etc. The largest material group that is used to manufacture piezoelectric sensors/devices is piezoceramics and piezopolymer, which because of their low weight and smaller size find enormous industrial applications.

Ceramics are another set of materials which are non-metallic in nature. These are produced from various inorganic compounds, such as oxides, nitrides, silicates and carbides. These are chemically resistant to most acids, alkalis and organic solvents and can withstand high temperatures. Whilemost metals weaken rapidly at higher temperatures of around 800 °C, the ceramics are in a position to retain their mechanical properties. Moreover, they are also good electrical insulators and possesses coefficient of thermal expansion very close to the metals that give them an edge over the other materials. Some ceramic materials, such as AlN and ZnO, also possess the piezoelectric effect, and these materials are classified as piezoceramics. Non-piezoelectric ceramics normally find applications in the making of tiles, bricks, pots, refractory bricks,other lining materials, etc.

There is one more type of materials that exhibits entirely different behavior in comparison to the metals, non-metals or ceramics. These are classified as a new group called as polymeric materials. Unlike the metals, these are composed of millions of repeating units (called as monomers) which are linked together to form long chains of the molecules (mostly of carbon atoms). These chains are also tangled or cross-linked together to give a physical form to the material. Monomers are basically a type of organic materials in which the carbon atoms are the common element that chemically combined with other elements such as carbon, oxygen, hydrogen and also with the non-metallic atoms (such as nitrogen).

Polymers exist naturally—known as biopolymer or natural polymeric materials. Amber, cellulose, DNA, hemp, protein, rubber, silk, shellac and wool are some examples of the natural polymeric materials. It is worth to note that the cellulose is the most abundant organic polymer on Earth. About 90% of the cotton fibers are cellulose content; whereas, in case of the wood, it is about 40–50%. It is also possible to design and manufacture the polymers in factories for some specific purpose and characteristics; such types of polymers are known as man-made or synthetic polymers. The list of the man-made polymers includes bakelite (phenol formaldehyde resin), polyethylene, polypropylene, polystyrene, polyvinyl chloride, Synthetic rubber, neoprene, nylon, polyacrylonitrile, silicon, etc. Many items of common use in our daily life are produced using these polymers. Silicon, for example, is extensively used as a waterproof sealant in many construction related activities. However, unlike some ceramics, the polymers do not possess the piezoelectric effect. But they still find many industrial applications and used extensively in comparison to the ceramics, etc. just because of their other important key characteristics, such as better biocompatibility, biodegradability and low production cost.

Biomaterials unlike the other materials are the special group of substances—natural or man-made, alive or lifeless—that possesses unique ability to interact among each other or with various biological systems to co-perform, augment or replace their natural function in the body. Additionally, they undergo almost zero volume change during their loading due to their Poisson ratio very near to 0.5; the For instance, bone implant materials are often designed to promote bone growth while dissolving into surrounding body fluid as they grow along with the natural growth of bone. Thus, for a material to perform as biomaterial, these must have a good biocompatibility, good strength and dissolution rates. These materials find enormous applications in the medical field, such as joint replacements, Bone plates, Intraocular lenses, bone cement, dental implants, stents, heart valves, etc.

1.3 Material Response

Materials depending upon their type respond differently, even to the same set of a physical disturbance, which can be an external force (mechanical loading) and/or temperature change (thermal loading) or any other cause that tends to either elongate or shorten the substance/object made up of that material. If the applied force increases the dimensions of the substance/object (may be its length or width or thickness or all), it is called as a tension force; otherwise, it is a compression force. This means when some physical disturbance giving a 'tension effect' (e.g., increase of temperature, tensile force, etc.) is applied to any material, it tends to pull its constituent units (may be atoms, molecules, crystals, aggregates, matrix, etc.) apart, away from each other. The consequence of this process is a reduction in the force of attraction that exists between the constituent units of the material—the more distance they move away from each other the smaller will be the force of attraction between them. On the contrary, the compression force moves them closer to each other, thereby leading to the increase in the bonding and/or friction and mechanical interlock. So, depending upon these interactions going on within the system and their cumulative outcome, the materials can respond in any one of the three possible ways, namely (1) elastic, (2) elasto-plastic and (3) plastic.

1.3.1 Elastic Response

Some materials respond elastically, meaning thereby that they will go on deforming under the applied loading conditions and follow a straight line load–deformation response before they finally fracture. The line OA in Fig. 1.10 is a typical load–deformation response that will represent an elastic response being exhibited by some materials. The key point is that if the load increments and the resulting deformations exhibited by the material are plotted together and we will get an exactly same curve from the loading (see 'OA' in Fig. 1.10) and unloading cycle (see 'AO' in Fig. 1.10) of the loading; such type of the material response is termed as elastic. Basically, it indicates the ability of a material to return to its previous shape after the load is released or removed. For instance, all brittle materials (like, ceramics, bricks, stones, rocks) exhibit an elastic response (line OA only, they will

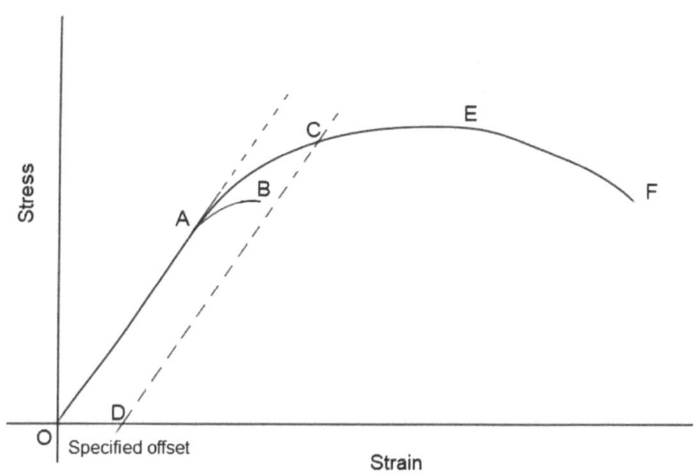

Fig. 1.10 Typical stress–strain response curve of a material under a tension force, applied axially. *Source* Author

fracture suddenly on reaching their ultimate load at the point **B**) before their sudden failure at the peak load (indicated by the point **B** on the response curve).

Usually, all ductile materials below their yield stress exhibit an elastic response. For some materials (like, the steel) exhibiting ductile response, sometime an imaginary point is used to indicate the onset of plastic response; it is indicated by specifying an offset (= OD, shown in Fig. 1.10). The point **C** on the response curve located by plotting a line parallel to the initial elastic curve (OA) gives the proof stress (an equivalent of the yield stress) for such type of materials. The point **E** gives the peak value of the material resistance, called as ultimate strength, and the point **F** on the curve quantifies the failure load of the material.

The prevalence of a lower stress level (corresponding to which materials normally exhibit an elastic response) does not cause any 'dislocation' [see, next sub-section] or loss of the bonding, which do not cause any irrecoverable damage to the material microstructure and as a result, the internal forces of attraction acting among different constituent particles/units of the material pull them back to their original locations on the removal of the loading. This process leads to the recovery of the original shape and size of the body.

1.3.2 Elasto-Plastic Response

When some loading is applied to an object, it will cause the object to behave in an elastic manner, wherein each increment of the applied load is accompanied by a proportional change in the deformation (it may be an elongation or shortening depending upon the load type: tension or compression, respectively). When the load is removed, the object returns to its original size. However, when the body is loaded beyond its elastic point (the point A shown on the curve, see Fig. 1.10), and once it exceeds a threshold—normally, its yield stress 'known as yield point'—the deformation increases more rapidly than that were observed in the elastic region, albeit it will happen almost at a near constant value of the load magnitude normally applied at the end of the elastic phase. Such type of the load–deformation response is known as an elasto-plastic response. A typical response is depicted in Fig. 1.10. It is important to note that after the loading, beyond the yield point, the material no longer unloads along the loading path (OAC) exhibited earlier by the specimen during an elasto-plastic response. Rather, it follows the path shown by a straight line CD, usually along a line drawn parallel to the initial loading line OA. Unlike an elastic response, there remains a permanent deformation/offset strain in the object even after the removal of the load, e.g., a metal rod under a tensile force (more than the load corresponding to its yield stress) will have a permanent change in its length and if the load level was kept less than the yield stress, then it will return to its original length after the removal of the load. And, if the body is loaded again, it will do so by loading along the line DC (with D as an origin in place of the point O) and repeat the same phenomenon in the case of any unloading done later, albeit with some new unloading line, maybe EG, and so on till the failure at the point F.

1.3.3 Plastic Response

In this type of the material response, materials having a negligible elastic range shift directly to their plastic range and go on deforming till its failure. For instance, the chewed bubble-gum plastically deforms enormously before finally breaking into two parts. Similarly, clay (in the moist state) gets permanently deformed when it is loaded either in tension or compression. A typical load–deformation curve for this type of a material response is shown in Fig. 1.6c. It depicts a very short elastic range being exhibited by any such type of materials and they enter their plastic range quickly and go on deforming over a very long range of the deformations. It is interesting to note that in the routine loading range (called as service load), no material is allowed to behave plastically, but in the case of some extreme loading conditions (which are very rare!), they serve as a good energy dissipater and help the engineers to optimize the member sizes by using the reserve material strength beyond their yield strength.

Almost all the metals and the specially designed composite materials (e.g., suitably reinforced concrete) undergo either an elasto-plastic or the plastic type of the material response; usually a consequence of *dislocations* which causes the atoms or group of atoms in the metals lying in one plane to break their bond, and again, develop new bonds with a new group of atoms on the other side of the slip plane after their dislocation under the effect of loading. This process is schematically shown in Fig. 1.11; the atoms across an imaginary slip line in some metal sample are shown using bubbles, typically numbered as 1, 2, 3, 4… Under the action of force (F), the atoms with an initial stable configuration of 1 and 5, 2 and 6, etc. start moving away from each other and gets relocated to some new configuration, e.g., 1 joins new atom, marked as 4 from its earlier attachment with the atom marked as 5; and the atom marked 2 gets new bonding with the atom 4, and so on.

This dislocation (in the metals) and the internal redistribution of the stresses when one of the material components in a composite gets exhausted on reaching its plastic stage to another material component is the 'carrier' of the plastic deformations being exhibited by them once these are loaded beyond their yield stress value. The result is that they are able to sustain the loading, without any failure,

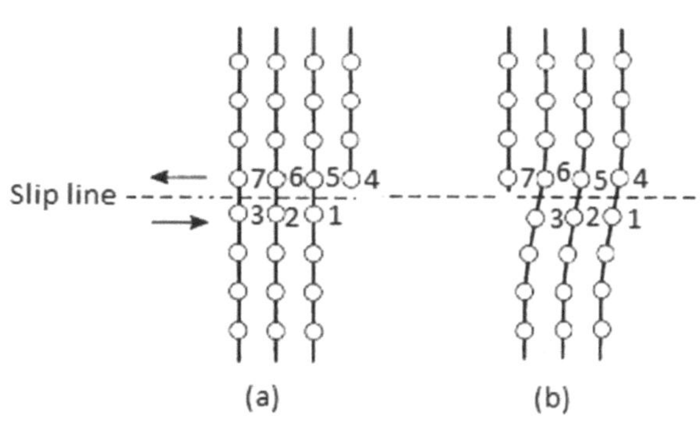

Fig. 1.11 Schematic presentation of the dislocation phenomenon: a typical stress–strain response curve of a material under a tension force applied axially. *Source* Author

over a long range of the deformations; the failure usually occurs when the material is no longer in the position to accommodate any more dislocations and/or redistribute the internal stresses safely.

1.4 Material Properties

All the materials whether natural or man-made possess a unique set of attributes that can be of some help to distinguish their different types and classify them into various categories. These material attributes that we can observe and measure using some set of procedures are known as their properties. A set of the properties that depend upon the quantity of material being tested are known as extensive properties. Some examples of such attributes are masses of the material and its volume —more the mass or volume; the more will be its value. There are many other material attributes—called as intensive properties—whose value does not depend on the amount of the substance for which we are conducting the measurements. For instance, the temperature of an object in the state of thermal equilibrium is the same as the temperature of any part of it. If the object is further sub-divided into different parts, the temperature of the each part will come out to be exactly same. Other examples can be material density, its load carrying capacity (strength), thermal conductivity, melting and boiling points, surface tension, pressure, etc. It is worth to note that many times, the property of material may be of some constant value or it can be a function of one or more independent variables, such as temperature. For instance, the strength of metals gets affected with the change in the temperature; normally it reduces when the metals are heated beyond some value or cooled below some temperature value. Similar is the case for other properties like surface tension, electric conductance, etc. Further, there are many materials for which properties vary to some degree according to the direction in which these are tested or measured—such a condition is referred to as anisotropy. Therefore, in addition to the knowledge of a list of different materials, their response is also of equal importance and, it plays an important role while selecting the most appropriate materials for a given set of field/design constraints.

Material properties, whether extensive or intensive, can be measured by conducting tests on their samples. The tests are usually conducted under a pre-defined set of standard controlled laboratory conditions. The nature of the test and the way it is required to be conducted all depends upon the material property being measured. The loading conditions during the testing operations can be mechanical or thermal or electrical/magnetic, etc. and accordingly, the response in term of the required attribute is observed and recorded. Generally, these material attributes are clubbed together under a unique head describing their response to a specific type of the loading condition, namely: (1) Mechanical, (2) thermal, (3) acoustical, (4) electrical, (5) magnetic and (6) chemical properties. Out of these, the properties that are used as a basis of the material classification and their subsequent selection in the context of civil engineering domain are discussed below:

1.4.1 Mechanical Properties

Whenever some material is loaded by applying a set of forces, e.g., two forces of equal magnitude, but acting on the opposite faces of the material/object either toward each other or away from each other, the material specimen deforms before its failure. The extent of the deformations depends upon the type of material being loaded; the brittle materials fail suddenly while the ductile ones go on resisting the applied load and deforming over a large range of the loading before they finally fail. During this process, the material deforms along the loading axis as well as in the direction normal to it (Think, why?). The possible observable key points during this loading process can be the value of maximum load taken by the material before its failure, the maximum value of the deformation (both along the loading axis and that normal to it), the load value corresponding to the onset of the yielding and the breakage/crushing point, the profile of the load–deformation points if plotted on a graph, etc. Accordingly, the following are the major key mechanical properties that help to characterize the materials on the basis of their mechanical behavior:

Strength: It is the ability of materials to sustain force without any form of failure; the stronger the material, the greater the force it can withstand just before its failure. All materials find their structural applications only because of this property as it determines their capacity to support the applied load and safely transfer it to some other point of interest. It is an intensive material property and is calculated by dividing the maximum load (P) carried by a material sample just before its failure by its cross-sectional area (A)—always taken normal to the loading axis, (= P/A). It is expressed as a ratio of the force units (N, kg, kN) and the area units (mm^2, cm^2, m^2), e.g., N/mm^2 (= 1 MPa) and is usually denoted by Greek alphabet *alpha* (σ). Each material possesses a unique value of the strength (σ), albeit it will be different if we test the material under a tension load or a compression load or shear load or bending action. The internal microstructure of a material plays an important role in deciding its strength properties. Normally, a dense packing of the constituent particles in the material makes it stronger than the material having a porous microstructure; it happens because of the stronger internal bonding between these particles when these are closely packed than it would be otherwise. For instance, the plain concrete is very strong if it is tested under a compression type of loading but its load carrying capacity reduced to about one-tenth of its compressive strength if the nature of the loading changes to the tension. Though the material in this case remains the same, its response will change with the change of the nature of the loading; the compression type of the force brings the particles closer while these starts moving away from each other under a tension force; thereby affecting the material response, which entirely depends upon the extent to which different stress transfer mechanisms are mobilized in the material. Similarly, the plain concrete has very low value of the bending resistance; it suddenly fractures into two segments once loaded continuously under a three-point loading, etc. (check out the possible reasons?).

If the strength of any material is known (from some experimental test data, etc.), then we can very easily determine the cross-sectional area of an object needed to support some quantity of the axial load. For example, a concrete sample has a compression strength (σ) of 5 MPa (say, it is the safe value), and we have to place a load (P) of 12,500 N over it. From this set of the data, the area of the concrete sample (A) needed to support this load can be calculated as $P/\sigma = 12{,}500/5 = 2500\ mm^2$. This area can be provided by suitably selecting dimensions of the specimen; it can be 50 mm × 50 mm, or 25 mm × 100 mm, or round specimen of 56.44 mm diameter. This way, there exist a number of options/cross-sectional shapes to transfer the given load through the material. Engineers have to select the most suitable shape/size for a given material type that meet their other design constraints also.

Stiffness: The materials deform during their loading. This process goes on till their failure, irrespective of the material type whether brittle or ductile. Materials resist the deformations depending upon their composition and the internal microstructure. There exists a unique load–deflection curve for every type of material, which describes their response to the applied load. The initial slope of this curve defines the material stiffness, k (= $\delta P/\Delta$); it can be calculated by dividing the load increments (δP) to their corresponding change in the deformations (Δ) being exhibited by the material sample during the loading process. In other words, the stiffness (k) of any material is a measure of the magnitude of the force needed to deform it by unity (= P/1). A high value of the stiffness indicates that the material under the investigations is very stiff and it will need a large magnitude of the force to deform it and vice versa, for instance, if we have a wooden log placed horizontally over the two objects at its end (assuming, each one possesses different value of the stiffness) and some weight is hanging at the center of the log. Just guess, what will be the position of the wooden log after its loading? It will not remain horizontal. Guess, why? The reason lies at the stiffness of the resting points over which the wooden log is placed at its ends. It is not the same, thereby leading to more deformation at the softer-end than the end with a higher value of the stiffness. Can we keep the wooden log in its initial horizontal position even after applying the load at its midpoint? Think, how? One possible answer will be to use the resting points/objects having the identical value of the stiffness. Then, one question will arise, what should be the stiffness of the resting points (objects) to limit the deformation there at 5 mm (say, any given constraint)? It can be easily managed by suitably selecting the materials at the resting points so that for the given value of material stiffness (k) and the applied loads (say, P) acting there translate to the said value of the deformations (= P/k). This way, this material property plays an important role in the selection of an appropriate material to achieve given set of design constraints. Because, it is the material strength and its stiffness that controls the behavior of any member while performing any desired structural action.

Elasticity and Plasticity: Elasticity is a material property which gives any object/member an ability to regain its initial shape and size once the load causing that deformation is removed. On the other hand, the capability of a material to deform permanently without any rupture or failure on removal of the load is called as plasticity; it is the just the opposite of elasticity. Each material possesses a unique value of the stress if it is stressed beyond that value of the stress, the resulting deformation becomes irrecoverable; the material is said to be entering its plastic state. And, if the stress level to which the material is loaded is kept less than that unique value, all deformations are recoverable and the stressed object will regain its shape and size on removal of the load. Normally, this unique value of the stress is called as material yield strength and it can be determined from the load–deformation response curve obtained during their testing.

In the elastic range, the deformations usually lead to either change in the change in the length or volume of the object. Depending upon the type of strain the material is undergoing during the loading process, different parameters (called as modulus) are used to quantify the material response, namely Young's modulus (denoted by **E**), bulk modulus (denoted by **K**), shear modulus (denoted by **G**). The first one expresses the relationship between the imposed stress and the linear strain, whereas the volumetric changes are shown by the bulk modulus. The shear modulus defines the relationship between the stress and the shear strain. Higher values of these moduli indicate need of a large force to deform the materials by any given amount. A typical value of these material parameters is tabulated in Table 1.1.

The tabulated data indicate that the shear modulus of solid materials is approximately 40% of their Young's modulus value and they (the metals) would exhibit about 30% lateral straining when these are deformed axially. The extremely small value of the shear modulus (G) of water is responsible for the liquidity it possesses; it simply moves sidewise whenever some stress is applied across any two sections. Similarly, the rubber changes its shape when pushed sidewise under the effect of some shear stress, which is negligible for most of the metals.

Toughness, resilience and hardness: Toughness is the ability of any material to deform, plastically, without any type of fracturing and is thus a measure of the amount of energy that it can absorb before its failure. The SI unit of the toughness is Joule per cubic meter (J/m^3). It can be calculated by finding the total area under the load–deformation response curve of the material. This is because of the fact that the area under the curve depicts the cumulative sum of the energy being stored in the material when it is continuously deformed under some load. It is worth to note that

Table 1.1 Moduli and Poisson's ratios for different materials

Material	Young's modulus, *E* (GPa)	Shear modulus, G (GPa)	Bulk modulus, *K* (GPa)	Poisson's ratio
Nickel	200	76	180	0.31
Mild steel	203	78	138	0.30
Copper	120	46	142	0.34
Aluminum	70	26	76	0.35
Rubber	14.9×10^{-4}	5×10^{-4}	1	0.49
Water	10^{-14}	10^{-14}	2.15	–

Source Author

for any material to be tough, it needs to be both strong and ductile; the higher the area of the curve the tougher would be the material. For instance, the brittle materials (like plain concrete) are strong, but they have a limited ductility; conversely, very ductile materials (like rubber) with low strengths are also not tough. Therefore, for a material to be tough, it must have a capability to withstand both high stresses as well as high strains.

It is important to recall that the strength indicates how much force the material can support, while toughness indicates how much energy a material can absorb just before its failure. The internal microstructure of the material is again playing its role in deciding the toughness. The most important part of a material to be tough is that it must be able to continuously switch the five major stress transfer mechanisms, listed in the previous section, from one to another without fail; the easiness with which it can do, the tougher the material will be. Out of all materials, the composites are the only type that can be designed to have a high strength as well as any desired value of the toughness (usually, high), just by playing with their microstructure, for example, by mixing an appropriate dose of short steel fibers in the concrete during its production stage make it stronger as well as tough. The reason is simple: the plain concrete is brittle in nature, so, as and when the applied load exceeds its threshold strain value the concrete fractures. However, the uniform presence of the steel fibers in the concrete will not allow it to respond that way. After the loss of the internal bonding (between different constituent units/particles) within the concrete mass, it continues to support the load by means of internal friction and mechanical interlocking being mobilizing along the uneven fractured surfaces along the slip/failure planes and subsequently, the bridging action of all those fibers which are crossing the fractured surfaces play their role in making it tough. Generally, all structural components that have to support a heavy static load as well as impact loading are proportioned to be tough as well as strong and the composites (e.g., reinforced materials) are the most suitable materials for this purpose.

On the other hand, the resilience of a material is the ability of a material or objects to spring back elastically into its initial shape/form; thus, it is a measure of the capacity of the materials to absorb energy elastically; whereas, the toughness is the total energy absorbed by any material just before its failure, which includes the elastic as well as the plastic range of the load–deformation curve. Therefore, by considering the area of the load–deformation response of a material up to its elastic range, the resilience can be calculated whereas the total area shows its toughness. Springs, shock absorbers, etc. are usually designed on the basis of the material resilience because the purpose of their use is to bounce back quickly to their original form, repeatedly, without failing. As the material is being loaded within its elastic range, the chances of permanently damaging it are very, very low if the springs are designed on the basis of the resilience in comparison to the material toughness.

Hardness is the property of a material due to which it offers resistance to penetration and scratching. Hard materials resist wear and scratches. Diamond is the hardest material.

1.4.2 Thermal Properties

Like mechanical loading, each material also responds differently to the thermal loading. The internal composition of materials decides their response; some materials expand more if we heat them up; while there are many others which do little, and sometimes, it may be insignificant and is usually ignored. Some materials remain hot for a longer period and others lose their heat quickly. Some materials soften easily at lower temperatures while there are many which needs to be heated to comparatively higher temperatures to reach their melt point. This mainly happens because of a change in the average molecular kinetic energy of a substance during the heating process. Normally, it increases when any material is heated and the effect is that its molecules start vibrating and moving more and more, which often leads to a greater average separation between them; thereby affecting the response of the materials. Based upon their response, a number of properties have been identified. Some of them which are very relevant to the civil engineering profession are given below:

Coefficient of thermal expansion: The changes in the dimensions/size of a body caused by a thermal loading are expressed by means of coefficient of thermal expansion; it is a measure of the fractional change in size that occurs in the body per degree change in the temperature at a constant pressure. Thermal expansion generally decreases with increasing bond energy, which also has an effect on the melting point of solids; therefore, generally high melting point materials are more likely to have lower thermal expansion. Several types of coefficients are in use depending upon the type of their applications and which dimensions are considered important in that application; it can be—linear, area and volumetric. The coefficients play an important role in the design process. For instance, both the plain concrete and ceramics are brittle materials and uneven temperature may cause their uneven expansion, which leads to the development of thermal stresses in the sample and finally leads to (undesirable) fracture. Therefore, if two materials with different coefficients of thermal expansion (say, its linear value) are used to produce some composite material, it may pose problems to maintain the desired composite action if its surrounding temperature is raised (check out, why?). Similarly, we will face problems when two materials of unequal coefficient of thermal expansions are joined together. The material with a higher value will undergo higher expansion than the second material. This process results in loosening of joints or change in the alignment, etc. of the complete member length. It is therefore all advisable to go for materials that possess an almost similar value of the expansion coefficients when it is not possible or is practically difficult to avoid their use in any specific application.

Specific heat: It is the amount of the heat energy required to raise the temperature of one unit mass of a substance by unity. It can also be interpreted as the amount of heat energy stored in the body (having a unit mass) if its temperature is raised by unity. And, this stored energy is then gradually radiated back to the surrounding to achieve a state of the thermal equilibrium with the surrounding environment. (Explore the ramification of this process in the context of buildings?) The SI unit for the specific heat is Joule per Kelvin per kilogram (J/K/kg). Since an increment of temperature of one degree Celsius (°C) is the same as an increment of one Kelvin; so accordingly, it is also sometimes expressed as Joule per degree Celsius per kilogram (J/°C/kg).

Generally, the matter in its gaseous state will have the highest value of specific heat among all its other forms. It can be even infinitely large when some material/object is undergoing a phase transition, such as melting or boiling, because the heat in that case will be used to change the state rather than raising its temperature. For instance, liquid water has one of the highest specific heats among all common substances in engineering applications. It is 4182 J/(K kg) at 20 °C for the water in its liquid state which reduces, by about 50%, to 2094 J/(K kg) in its solid state (i.e., ice). This value is significantly higher if we compare it with a specific heat of some other commonly used solids, such as iron, granite, which is only 449 and 790 J/(K kg), respectively.

Thermal conductivity: It is the ability of the material to conduct heat. Heat transfer occurs at a lower rate in the materials of low thermal conductivity and vice versa. For instance, metals typically have a high thermal conductivity and are very efficient at conducting heat, while the opposite is true for insulating materials, like foam and wood. In various engineering applications, it is measured in terms of 'thermal conductance' which is defined as the quantity of the heat energy that passes per unit time through a body (of specified thickness) when its opposite faces are maintained at a temperature difference of one Kelvin (or one Celsius). The SI unit of this material property is Watt/Kelvin (W/K or W/°C). Gases generally have low thermal conductivity, among all forms of the matter; whereas the pure metals have comparatively higher thermal conductivity.

Flammability: It is the ease with which a combustible material can be ignited thereby leading to a fire. As such, this material property is an important design parameter to decide the type of material needed, especially in a case where fire resistance is the prime requirement or in all industrial processes that may produce combustible substances as a byproduct and also for all construction activities where it is used as such (e.g., wood) or is being stored for some other purpose. Special precautions are usually required for all such types of materials that have a high flammability. These measures may include the installation of fire sprinklers or storing them remotely from the possible sources of ignition. The flash point of the material is one of many parameters generally used to indicate its flammability.

The flash point is the lowest temperature at which there will be enough flammable vapors being liberated from the material into its surroundings to induce ignition in the form of flash under definite conditions of the test. Importantly, at this point, the fire will not last long; just a flash will appear for a fraction of a second. It is an empirical measurement rather than a fundamental physical parameter as its value varies with the type of equipments and the test protocol variations used in its determination, such as a temperature ramp rate (as in case of automated testers), the time allowed for the sample to equilibrate, the sample volume and whether the sample is stirred or not. For instance, bitumen is viscoelastic materials used in various construction activities, such as road construction, sealant for water proofing, etc. It did not have any sharply defined melting point; however, they gradually become softer and less viscous with the rise of temperature, and during this process, they release volatiles into the atmosphere which are prone to ignite fire in case of any mishandling, improper use at the site. Therefore, this parameter helps to know the safe mixing and application temperature values of a particular bitumen grade. Similarly, depending upon the site conditions, it assists in the selection of the most suitable material, especially in areas where fire sensitivity is of prime importance.

1.4.3 Acoustical Properties

The parameters associated with this property, such as the speed of sound, acoustical absorption, etc. describe the response of the material to the sound energy and helps engineers to choose most appropriate material required to control the acoustics of buildings, seminar halls or auditoriums, etc.

Speed of sound: Materials depending upon their microstructure transfer the sound energy differently through them. Some materials do it rapidly, such as solids; while there are many others which respond slowly and do at a slower pace. For instance, sound travels at 343 m/s in the air (gas); it travels at 1480 m/s in water (liquid) and at 5120 m/s in iron (solid). Amazingly, it can be as high as 12,000 m/s in exceptionally stiff material like diamond. This happens because of the fact that the sound energy travel in the form of compression waves in solids and its microstructure which defines its compressibility, shear modulus and density controls the transmission of this energy; the finer the microstructure of a material denser it will be, which results in a fast transmission of the sound energy. In engineering applications, this parameter helps to check the quality of material used at sites; the materials produced with a poor quality usually lead to a porous microstructure thus lowering the speed of the sound waves traveling through it than it should be otherwise. It can also be used to detect the internal cracking in materials which normally delays the efficient transmission of the energy. The ultrasound pulse velocity (UPV) equipment is very common equipment that works on this principle and is used to check these two types of the field problems. Alternatively, materials can be intentionally produced porous to reduce the sound transmission when the purpose of their use is to minimize the disturbances across different areas of the building.

Acoustical absorption: It refers to the process by which any material takes in sound energy when it encountered the sound waves. The amount of the energy it is taking in during this process is called as absorbed energy. Generally, some part of the absorbed energy is transformed into heat, while its other part is transmitted through the absorbing body; more the dissipation of the energy the better the material will be. This property plays an important role in deciding suitable materials for doing soundproofing of assembly/seminar halls, auditorium, etc. where the main aim is to absorb as much sound energy as possible, and converting the rest part into heat or transmitting it away from a certain location. In general, soft and porous materials serve as good acoustic insulators—absorbing most of the sound energy, whereas dense, hard and impenetrable materials reflect the most and therefore the materials with such characteristics should be avoided if the purpose of their use is sound proofing.

Acoustic transmission: It is the transmission of the sound waves through and between the different materials. The degree to which it is transferred between two materials depends on how well their acoustical impedances match during the process. The acoustic impedance of a system is a measure of the opposition that it offers to the acoustic flow produced by an acoustic pressure applied to the system. The SI unit of acoustic impedance is the Pascal second per cubic meter (Pa s/m^3).

1.4.4 Electrical Properties

The parameters associated with this property, such as conductivity, resistivity, etc. describe the response of the material to an electric field. Electric current refers to a flow of electrons in materials. Some materials, such as metals conduct electric current easily and faster than others like ceramic, rubber, etc. This happens because there are many electron energy levels in the metals nearer to the 'Fermi level,' which enables the availability of a large pool of the electrons for a free movement under the influence of external electric field, thereby giving them high electric conductivity. In civil engineering applications, electric resistivity/conductivity and piezoelectric constants are commonly used in comparisons to the other properties, like capacitance, dielectric constant, permittivity, pyroelectricity, electric susceptibility, electrostriction, etc. and these two properties are described below:

Electrical resistivity and conductivity: These are the fundamental property of a material that quantifies how strongly it resists or conducts the electric current, respectively. Usually, these two parameters are reciprocal of each other; a high resistivity of a material means it is of low conductivity and vice versa. A low resistivity indicates a material that readily allows electric current. The SI unit of electrical resistivity is the ohm-meter (Ω m). For example, if a 1 m × 1 m × 1 m solid cube of a material has sheet contacts on two opposite faces, and the resistance between these contacts is 1 Ω, then the resistivity of the material is 1 Ω m.

Piezoelectric constants: There are two types of the constants, namely the piezoelectric charge constant and piezovoltage constant. These are mainly classified on the basis of the material response to an electric field or some mechanical field. This is because any material exhibiting piezoelectricity responds to an electric field in the form of mechanical straining and also, in the form of electric charge that will appear on its opposite faces when it is subjected to some stress field. Piezoelectric charge constant is the polarization generated per unit of mechanical stress applied to a piezoelectric material and the piezovoltage constant is the mechanical strain experienced by a piezoelectric material per unit of electric field applied. The high value of the constant indicates that the material will undergo relatively higher strain when some electric field is applied across it. Because the strain induced in a piezoelectric material by an applied electric field is the product of the value of the electric field and the piezoelectric constant of that material. Thus, it is an important indicator of a material's suitability for strain-dependent applications, such as sensors, actuators, etc.

1.4.5 Magnetism

Some materials, such as iron, exhibit unique response when placed in some magnetic field. This happens because of presence of tiny magnetic fields in the materials at their atomic levels being produced by the orbital and spin motion of electrons, which led to the creation of tiny atomic current loops and the corresponding magnetic fields. When such materials are placed in some external magnetic field, these current loops tend to align in such a way so as to oppose the applied field; this leads to the generation of a force of attraction/repulsion depending upon the placement. This helps the material to align in the direction of the external magnetic field. This process helps engineers to control the mechanical behavior of composite which contains ferrous materials like steel fiber reinforced concrete specimens. If the concrete containing steel fibers is poured in the mold placed in some magnetic field, the fibers tend to align themselves in the direction of the field thus controlling the response of the concrete containing steel fibers. Generally, the fibers aligned along the sample length are most effective in carrying the external load. For steel fibers, it is relatively easy to manipulate the direction of the fibers in the concrete mix by adjusting the external magnetic field.

It is amazing to know that all available matter possesses magnetic properties; the only thing is that some materials are much more magnetic than others. The magnetic behavior of materials, therefore, can be classified accordingly into the following three major groups, namely (1) diamagnetism, (2) para magnetism, and (3) ferromagnetism. Materials in the first two groups are generally those that exhibit no collective magnetic interactions when placed in close vicinity to each other and are generally so weakly magnetic that they are considered as 'nonmagnetic'; while the third group materials exhibit magnetic behavior below a certain critical temperature, known as Curie temperature. In this group, one very well known and the common material that possesses the magnetic properties is Iron.

1.4.6 Chemical Properties

These describe the chemical response of materials when these are placed in contact with other materials under certain physical or thermal conditions. The simple process to illustrate this material aspect is the response of iron in the presence of water and air. It corrodes and chemically changes to new more stable substance. Similarly, when two dissimilar materials are placed in close proximity, they react and influence the overall response of the combined product. Some of the important properties that fall under this head as described below:

Corrosion resistance: It is a natural process that converts a refined metal into its relatively more chemically stable form, such as oxide, hydroxide, etc. The materials most resistant to corrosion are those for which corrosion is thermodynamically unfavorable. For instance, the corrosion products of the metals like gold or platinum tend to decompose spontaneously into the pure metal itself, which is why these elements can be found in the metallic form on Earth. There are many metals which, despite their slow reaction kinetics remain virtually corrosion free, e.g., zinc, magnesium and cadmium. While corrosion of these metals is continuous and ongoing, bit happens at a very slow rate and remains mostly unnoticeable. While for the iron, it is very rapid under favorable conditions and converts the iron to its oxides (see, Fig. 1.12). Interestingly, these products of corrosion are very voluminous and badly affect the materials placed in close proximity to the Iron. The very common example is spalling of concrete wherein the products of corrosion of embedded reinforcing bars tend to dislodge the concrete cover in the reinforced concrete member.

Hygroscopy: It is the phenomenon of attracting and holding water molecules either through absorption or adsorption from the surrounding environment, which usually occurs in the room temperature. The amount of the moisture held by any hygroscopic material is usually proportional to the relative humidity. This behavior of the material significantly affects the material selection process. For instance, we have to be very careful while using materials like sugar, honey, glycerol, ethanol, many fertilizer chemicals and salts like calcium chloride, Zinc chloride, Sodium hydroxide, sodium chloride, etc. Out of these listed materials, some like zinc chloride, Calcium chloride and sodium hydroxide are so hygroscopic that they readily dissolve in the water they absorb. So, when using such materials, the users need to be very careful while handling them or storing them.

(a) (b)

Fig. 1.12 Typical view of the corroded rebar embedded in concrete: **a** the voluminous products of the rusting, **b** which leads to the spalling and internal cracking in the concrete members. *Source* Author *pH*: The pH scale is logarithmic in nature and the value of the parameter inversely indicates the concentration of the hydrogen ions (OH^{-1}) present in the solution. It is the negative of the base 10 logarithm of the activity of the hydrogen ion in the solution. Acidic solutions have a lower pH, while basic solutions have a higher pH. At the room temperature (25 °C), the pure water is neither acidic nor basic and has a pH of 7. However, if the water is exposed to air it becomes mildly acidic because it absorbs the carbon dioxide from the air, which is then slowly converted into bicarbonate and hydrogen ions and thus lowering its pH value. This parameter has a significant role in regulating the corrosion potential of steel/iron reinforcing bars embedded in concrete members. Because, the bars start corroding the moment pH value of the paste surrounding them go below 9. It otherwise remains around 12 because of the presence of $Ca(OH)_2$. However, when CO_2 from the moist air find its ways into the concrete microstructure, it starts reducing the pH which triggers the corrosion of rebars.

Reactivity: Reactivity refers to the rate at which some material tends to undergo a chemical reaction. Thermodynamically, the reactions happen to bring the chemical composition of the materials back to their lower energy state; thereby, giving it a more stable state. In case of the pure compounds, their reactivity is governed by the physical properties of the material. Normally, materials having a higher specific surface area are more reactive than otherwise. This can be done easily by grinding them to a fine powder form which will have a more specific surface area than its coarser form. The reactivity of materials is also influenced by the presence of contaminants and crystalline form which help to trigger the chemical reactions. In crystalline compounds, this can also affect their reactivity. This property is very crucial to produce different types of cements. It also controls the behavior of different types of clays.

Revision Questions

Q1. What do you mean by material? Write down various forms of material that exist in nature and give some examples for each case that find applications in civil engineering.

Q2. Identify and enlist different types of structural material used in the construction of your home.

Q3. Enlist various desirable properties of structural and nonstructural materials.

Q4. List the five 'stress transfer mechanisms' that enable materials to resist externally applied loads?

Q5. Differentiate strength and stiffness. Illustrate the concept using suitable example.

Q6. How do the size and shape of aggregates used in the production of a composite material (like concrete) influences its strength properties?

Q7. The maximum permissible displacement for a concrete cube of 150 mm size is 2 mm. What should be the highest magnitude of vertical load that we can apply to the concrete block? The stiffness of concrete used in the production of the block is 23500 N/mm. And, check whether the concrete if subjected to that much loading will crush it or not? The compression strength of the concrete is 20 MPa.

Q8. What do you mean by stress–strain response of the material? Discuss the importance of this important material's behavior in engineering design.

Q9. How can we classify the materials if their stress–strain response is known from some experimental test data?

Q10. Differentiate strain-softening and strain-hardening response exhibited by materials. Explain in detail with the help of appropriate example.

Q11. The reaction of cement with water is exothermic in nature. How this aspect of material response does affect their strength properties? Discuss in detail.

Q12. Prepare a catalogue of structural materials that find their applications in various civil engineering projects. It must include:

- Name of the material and its type
- Photographs (depicting its texture, color, microstructure, etc.), and original sample
- Physical properties
- Chemical composition
- Engineering properties
- Mode of internal stress transfer
- Production/availability
- Application range
- The skill needed to use/produce
- Source of all this information used in the preparation of the catalogue.

Q13. Define toughness of the material. How does it differ from its strength? Explain with the help of a suitable example.

Q14. How does the stress–strain response of a material help to determine its toughness?

Q15. Is it possible to improve/enhance the toughness of a material? If yes, How? And if no, Why? Discuss in details.

Q16. Differentiate resilience and toughness of a material. Discuss the importance of each parameter.

Q17. How the strain-controlled loading does differs from the stress-controlled loading?

Q18. Can we calculate toughness of a material using the stress-controlled testing? If your answer is in yes, explain, how? And if no, why?

Q19. Enlist and describe various factors that control toughness and strength of materials. Which types of material have higher toughness; name any ten such types of materials.

Q20. Define characteristic strength. Why is it that engineers prefer this strength in the design?

Bibliography

1. Budinski K (1979) Engineering materials, properties and selection. Prentice-Hall, Englewood Cliffs, NJ
2. Charles JA, Crane FAA, Furness JAG (1987) Selection and use of engineering materials, 3rd edn. Butterworth-Heinemann, Oxford
3. Cottrell AH (1964) Mechanical properties of matter. Wiley, New York
4. Dieter GE (1991) Engineering design, a materials and processing approach, 2nd edn. McGraw-Hill, NY
5. Farag MM (1989) Selection of materials and manufacturing processes for engineering design. Prentice-Hall, Englewood Cliffs, NJ
6. Fontana MG, Greene ND (1967) Corrosion engineering. McGraw-Hill, New York
7. Fulay PP, Wright WJ, Askeland DR (2011) The science and engineering of materials. Cengage Learning, United States
8. Gordon JE (1978) Structures, or why things don't fall through the floor. Penguin Books, Harmondsworth
9. Lewis G (1990) Selection of engineering materials. Prentice-Hall, Englewood Cliffs, NJ
10. Blazynski TZ (1987) Materials at high strain rates. Springer, Netherlands
11. Tabor D (1978) Properties of matter. Penguin Books, London

Chapter 2
Material Behavior

Materials depending upon their type respond differently to physical disturbances, which can be an external force (mechanical loading) and/or temperature change (thermal loading) or any other cause that tends to either elongate or shorten the substance/object made up of any such material. The internal arrangement of the constituting blocks of materials (may be their atomic and/or lattice structure) plays an important role in dictating the response they would exhibit under the action of any such loading conditions. This arrangement of the materials is known as their microstructure; it is very unique for each and every category of the materials. If some material is able to retain the overall form of its microstructure by progressively shifting the mode of transfer of internal forces from one to another till its failure, it will be categorized as tough material; otherwise, it will be a brittle one. Irrespective of whether brittle or tough, they behave elastically in their initial range of the load–deformation response curve. Nevertheless, it is always possible to alter the response of materials and improve their mechanical properties. The following sections describe different forms of the atomic and microstructural arrangements that various materials possess and how these control their mechanical response. At the end of the chapter, a selection of textbooks and journal articles is listed, which might be helpful for the readers to deepen their understanding of these processes.

2.1 Material Microstructure

Materials (especially in their solid form) retain their shape/geometry as they are capable of transferring the stress from one point (of its application) to some other points (of interest) in them safely, albeit it occurs along with the deformations. The resulting deformations can be negligible or small in magnitude as most of the solids do, or these can be very large, generally uncontrollable, as exhibited by liquids or gases. In such a case, it will not be possible for the liquids or gaseous materials maintain their pre-assigned shape/size; rather they will take up the shape of the

H. Singh, *Structural Materials*, Materials Horizons: From Nature to Nanomaterials,
https://doi.org/10.1007/978-981-16-3211-2_2

container used to store them. In the absence of any container, they tend to move in the direction of the applied force created by the pressure difference. In doing so, either one or more of the five principle modes of stress transfer in materials; namely bonding; friction; mechanical interlock; surface contact and dowel action gets activated that help them either keep their pre-assigned geometrical dimensions, i.e., its shape and size, or cause them to move like a liquid or gasses. Normally, the first one of these is fundamental to almost all types of known materials, and it operates at the atomic and/or the molecular level in them. The second one is common for both solids and liquids; whereas the remaining modes get activated in the solids only depending upon their type and magnitude of the applied load.

At the atomic level, all forms of the matter that exist in the universe are composed of atoms. Overall, these are electrically neutral, but internally, all atoms have a set of charged particles, namely the electrons (negatively charged particles) orbiting the nucleus of an atom, which in itself is a positively charged entity containing protons and neutrons. The atomic number is used to indicate the number of protons that any atom can have in its nucleus. The number of electrons in atoms being always equal to the number of protons can be determined easily from the atomic number of any element, but each orbit or shell in any atom can hold only a fixed number of electrons depending upon their energy level; generally equals to $2n^2$, where n is the shell number. For instance, the first orbit of an atom can have only two electrons. Similarly, the second and the third shell can have 8 (= 2 + 6), 18 (= 2 + 6 + 10) electrons, respectively. The break-up shown in the bracket indicates sub-shell numbers (*s*, *p*, *d*, etc.) of various orbits; the electrons first fill up the lower energy level (e.g., *s*-level), then the remaining electrons go to the next levels (may be, *p* and then, *d* and so on). However, the ways in which the electrons in the outermost shell of atoms (known as valence electrons) interact among each other decide the type of an atomic arrangement that any element will have after the formation of the primary bonding, and this arrangement gives solids their unique massive shape and form.

These interactions may lead to the formation of an ionic bond, when the electrons from the outer shell of one atom move to another atom to achieve a state of stable configuration; the atoms which lose some of their electrons in the process become positively charged, whereas the receiver becomes negatively charged entity; it naturally leads to the creation of attractive forces between them. This occurs, for example, in common salt when sodium and chlorine join to form sodium chloride molecule (NaCl). First, sodium atoms (Na) oxidize and lose an electron to form the positively charged sodium ions (Na^+), whereas, chlorine atoms gain the electrons from the sodium atoms to form the negatively charged chloride ions ($C1^-$). Both ions being oppositely charged leads to strong electrostatic forces of attraction between them to give them a solid form that we normally see in the form of white crystalline particles of the common salt.

It may also result in the covalent bonding when two adjoining atoms in the metal/material matrix start sharing their electrons in the outer shell among each other to achieve a state of the stable configuration. The shared electrons in the two covalently joined atoms continue to switch between them as they move in their orbit thereby giving a stable structural form. Normally, this happens in case of the non-metals which have a high number of the valence electrons (generally, four or more), and they prefer to gain electrons, instead of losing them, in the chemical reactions; they very often prefer to form anions during this process. In a molecule formed from two or more non-metals, there are no elements willing to become cations, so the ionic bonds are not possible for such type of molecules. Instead, two non-metallic atoms can share valence electrons with each other which lead to the formation of the covalent bonds by means of electron sharing; the shared electrons become part of both atoms in the setup. The calcium silicate chain in case of the concrete is a fine example of covalently held molecular structures, wherein the chain consists of a number of stable Silicate Tetrahedron formed by a Silicon atom linked with four Oxygen atoms.

The third possibility is the formation of the metallic bonding wherein the loosely held electrons leave their parent atom and start moving freely among many atoms in the material. So, there is no option for the positively charged atoms (after they lose their electrons temporarily) but to keep the flock of the free electrons together among them to achieve a stable configuration. All metals posses a very regular and well-defined structured arrangement of atoms because of the presence of metallic bonding in them. These repeating structural units in the metals are called as unit cell. These cells lead to the formation of a lattice structure—a well-defined, very unique geometrical pattern for each and every type of the metal. A cluster of entangled set of lattice structure leads to the formation of grains. At the microlevel, therefore, all metals posses a large pool of mutually packed grains. The metal production process has a significant effect on the shape and type of the lattice structure, and the grain sizes that the metal would develop upon its cooling. The resultant microstructure of the metals plays an important role in dictating the response that these would exhibit under a set of externally applied loading conditions. Generally, all metals exhibit high conductance (both, electric and thermal) and strength properties because of the availability of a large pool of free electrons in them.

The physical arrangement of the atoms in molecules sometimes results in unbalanced electric charges on its opposite faces—positive charge on one of its faces and negative charge on the other. So, when two such molecules come near to each other, they have a natural tendency to join each other and attain a state of stable configuration. This type of binding among different molecules is called as secondary bonds. As valence electrons in atoms are not involved in the formation of the secondary bonds, this type of the bonding is much weaker than the primary bonds, but it contributes greatly in the formation of many materials. In polymers, for an instant, this type of the force of attraction plays an important role in binding various long chains of polymers together. The force of attraction being induced between different atoms and/or molecules is known as van der Waals forces. The dipole–dipole and the hydrogen bonding are examples of such interactions existing at the atomic or molecular level in many materials.

The dipole–dipole forces exist between the positively charged face of one polar molecule and the negative face of another nearby polar molecule. These types of forces are most dominant when such types of molecules are lying very nearer to each other. It may also activate when the electrons in some atoms (like hydrogen) move in the orbit around the nucleus. As it revolves around the nucleus, the position of an electron at any instance keeps that side of the atom negatively charged while the opposite side of the same atom becomes positively charged though temporarily. But this process is able to generate dipoles in the material, which becomes a source of keeping different constituent units intact. Unlike the van der Waals forces, the polarity in the case of the hydrogen bonding also exists because of the hydrogen atom that covalently joins with some other nearby electronegative atom, such as N, O or F, etc. The water molecule, for example, exists as dipolar because of asymmetry about one of its axis. The hydrogen atoms in the molecule are oriented at an angle of 105° with respect to each other while joining one atom of the oxygen atom. This arrangement of atoms in the molecule leads to polarity which keeps various other molecules in position and all exist in liquid form at the room temperature. However, with rising temperature these molecules tend to move away from each other with the weakening of van der Waals forces, and finally, these escapes as gas molecules in the air.

2.1.1 Metals and Non-metals

Metals and non-metals exist in a solid form because of the metallic bonding that keeps all the atoms together (But, some exceptions are there: Mercury and bromine, respectively, which exist as a liquid at the room temperature). The free flock of electrons continues to shift their positions among different atoms in them, thereby creating electrostatic forces of attraction, responsible for the solidness they posses. Some of their properties, e.g., elastic constants, are directly related to the extent of the metallic bonds between the atoms. Ideally, this type of the internal placement of atoms produces a very regular and well-structured arrangement of the repeating arrays called as unit cell. Each cell has a certain number of atoms held together in some unique configuration. The arrangement is so exact that it can be reduced down to one small piece that simply repeats over and over again in every direction. The mutual placement of these cells in the metals leads to the formation of a crystal or lattice structure—again a well-defined, very unique three-dimensional geometrical pattern for each and every type of the metal. Three different types of lattice structures have been identified that develop in the metals depending upon the process used in their production; namely (1) face-centered cubic, (2) body-centered cubic, and (3) hexagonal closed-packed. Figure 2.1 depicts a typical form of these three possibilities that generally develop in the metals. Because of this, almost all metals possess polycrystalline-multiphase microstructure. However, many metals and their alloys can also exist in more than one latticed structural form depending on the process used to manufacture them and also their chemical composition, but transitions can also be present in many cases where the metals can have a mix of the standard lattice structural forms.

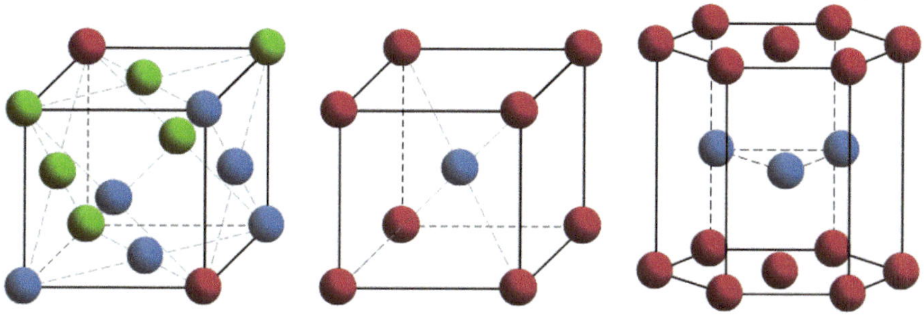

Fig. 2.1 Different types of the lattice structures that metals and non-metals develop depending upon the production process. The dots in the lattice denote the position of the atoms. *Source* Author

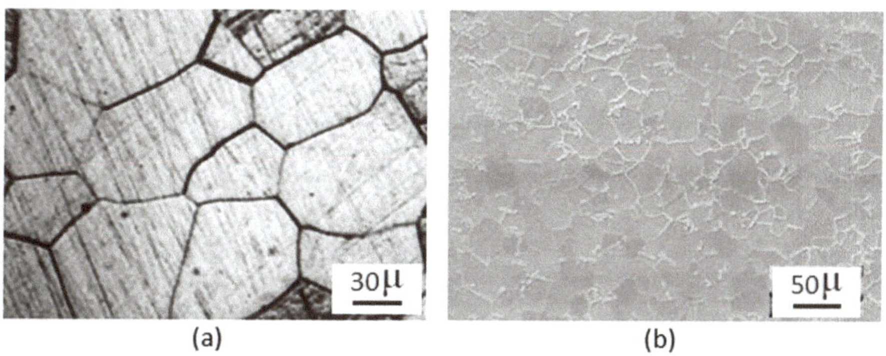

Fig. 2.2 Typical depiction of grains in the metal: **a** Grain boundaries evidenced by acid etching, **b** microscopic view (Image *Credit Edward Pleshakov*; *Source* https://commons.wikimedia.org/wiki/File:CrystalGrain.jpg)

If we move out of the crystal level of materials, the grains are the major influencing entities in the metals which control most of their mechanical behavior; these are nothing but an assembly/cluster of crystals present in the metals. A grain is a single crystal surrounded by other crystals of the same type (see Fig. 2.2) but with some other orientations. The grain size depends on the material processing techniques and the other heat treatments done during the production process, but it can be adjusted in a wide range by controlling different parameters being used in the production processes. The normal grain size in the metals varies between 1 and 1000 μm. Within each grain, the atoms are regularly arranged according to the basic crystal structure of the metal, but a variety of imperfections, termed as crystal defects, may also occur during the heat treatments given at the time of the reduction or casting process. Depending upon their form, four different types of these defects are classified; namely point defects (vacancies), line defects (dislocations), planar defects (twin boundaries) and volume defects (voids or cavities).

These defects are the sources of weakness in the metals which otherwise should be as strong as the force required to move the unit cells away from each other. But the presence of these defects in the metals significantly reduces their load carrying capacity. Among all defects, the dislocations are of particular interest as the plastic deformation that occurs in the metals at the large stress level are the nothing but the internal movement of these grains along the line defects, whereas the point and volume defects act as trigger points to initiate the dislocation in the metals. Restricting this type of movement can be of some help if someone wants to improve strength and other mechanical properties of the metals, and this procedure is very basic to all strengthening mechanisms generally adopted in the practice.

The ease with which the dislocations in the metals move under the action of applied stress decides their strength. The grain size has a significant influence on the strength exhibited by the metals. The smaller the grain size of the metal, the more will be its strength. Normally, a linear relationship exists between the yield strength of a metal and the reciprocal value of the square root of its grain size, normally expressed as the diameter of a grain. The metal having fine-grained texture is stronger than the one that is coarse grained, since the former has a greater total grain boundary area to obstruct dislocation movement, and accordingly, more energy will be needed to dislocate the same. Metals also contain grains of various shapes and sizes; some of them may be as large as ten times the smaller one present in the metals. These non-uniform sizes and shapes of the grains and their relative irregular placement in the metals also affect their strength. If a dislocation passes from one grain to another in such type of a network, it will have to change its direction frequently, which normally need higher efforts to move the grains by the same amount in comparison to the metals which posses uniform grain sizes, evenly place throughout the body of the metals.

Small grain size also creates preferred planes (known as slip planes) in the lattice where the atoms (in the form of dislocations) are free to move across each other. In case, the direction of the applied stress coincides with a slip plane, the dislocations can move easily. If the applied stress is acting perpendicular to the slip plane, it would be extremely difficult for the dislocations to move. Each grain, therefore, is weaker in certain directions than in others. However, with many grains oriented in random directions, the microscopic directionality of the strength would tend to be averaged out, which finally leads to equal metal strength in all possible directions.

2.1.2 Polymers

Polymeric materials composed of millions of repeating units (called monomers) which are linked together, mainly through covalent bonds, to make long chains of the molecules. These chains are further tangled or cross-linked to give a physical form to the material. Generally, the extent of the cross-linking and branching in the polymeric chains decides their stiffness and strength. The more 'lumpy' and branched the polymer, the less dense and less crystalline it will be. Carbon is the

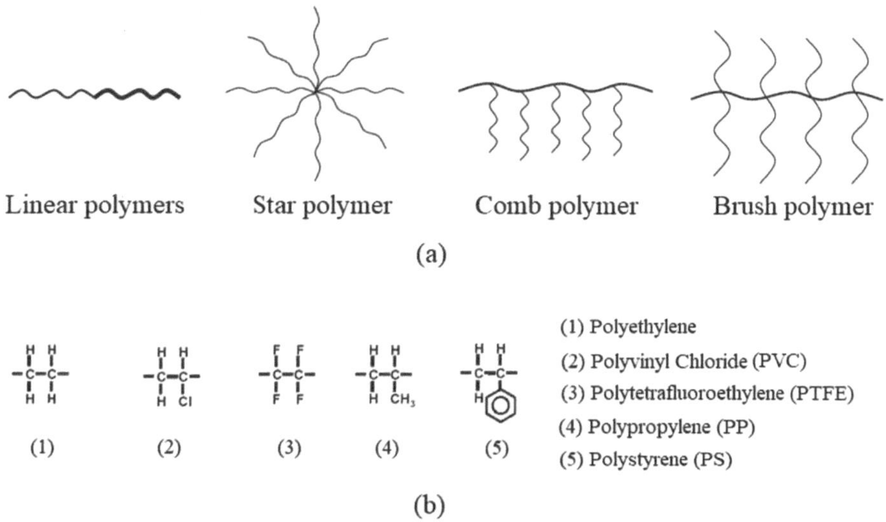

Fig. 2.3 Typical representation of molecular arrangement of polymeric materials: **a** different polymeric layouts, **b** some of the monomers. *Source* Author

most common element of any polymeric chain, which chemically joins with many other non-metallic elements to form long chains of monomers. This becomes possible as the carbon atom has only four valence electrons which it can share with the other atoms easily by means of the covalent bonding. Long and strong chains made of thousands of carbon atoms so forms are a backbone of a polymer. Figure 2.3 depicts the molecular arrangement that normally exists in polymeric materials. It can exist naturally—known as biopolymers or natural polymeric materials as well as it can be produced artificially. Amber, cellulose, DNA, protein, rubber, silk, shellac and wool are some examples of the natural polymeric materials, whereas polyethylene, polyvinyal chloride, polytetrafluoroethylene, polystyrene, polypropylene, silicon, etc. are some of the man-made polymers.

Unlike the metals, polymers have low densities; they do not conduct electricity nor do they absorb light. The entirely different chemical structures of various polymeric materials make it possible to have such contrasting properties in comparison to the metals. Because of covalent bonding, the valence electrons in them move around a molecule may not be symmetrically distributed. The non-metallic elements, such as nitrogen, oxygen, fluorine and chlorine, with which the carbon forms polymeric chains tend to shift shared electrons away from carbon and hydrogen. This shifting leads to polarization of the monomers with negative charge on one of its faces and the positive charge on the other side the molecule (with carbon and hydrogen). Its positive sections are attracted and joined to the negative side of neighboring polymers. Because of the presence of such type of internal bonding among different polymeric chains, these posses low densities and are flammable in nature with a low boiling point. Depending upon the type of the internal arrangement of the polymeric chains and their mutual bonding among each other, they can be categorized into three different classes, namely fibers, thermoplastic, elastomer.

Fibers exist in the form of thread that are usually flexible and can be tightly woven without breaking. A typical molecule, in this case, composed of long carbon chains with few or no branches. Silk, nylon, polyester and cotton are a few examples of fibers. When many fibers are put up in parallel, they form long strands which are much stronger than the individual fiber. Their chemical composition and the internal intermolecular forces lead to an ordered (rather than crystalline) physical form, an arrangement which gives them very distinct physical properties. They are axially very strong in tension. However, if you pull on both ends of a strand, it will not stretch very much as the chains are already aligned during their production.

The thermoplastic is a class of polymeric materials that are very flexible at high temperatures, but remain stiff under ambient conditions; however, their shape can be changed easily without needing much force. The internal arrangement of various polymeric chains in this class of material makes that possible. Various molecular chains in such type of materials are very random in their orientation and are held together by means of weak secondary bonds. At the low temperatures, the polymeric chains will not slide past each other because they do not posses enough energy to overcome the secondary bonds and consequentially, they remain stiff. As a result, the products made from such materials are bendable and become soft as the temperature rises. Small amount of force can make the chains slide around, thus changing the shape of objects made up from such materials. Small molecules called plasticizers are often mixed into the thermoplastics. They reduce the number of secondary bonds that can develop between the long polymer chains, making it easier to modify the shape. The glass transition temperature of polymeric materials reduces by addition of chemicals called as plasticizers, which finally evaporate over a period of time (a process that speed up at high temperatures), thereby leading to increase in the material brittleness. The elastomers on the other hand are highly stretchable polymers; these can be extended in their length manifold just by pulling their opposite ends. Rubber is a common example of such type of polymers. The molecular structure includes chains that prefer being tangled. If the strands made from such materials are stretched, a restoring force being generated through this entanglement pulls them back into their initial tangled position.

2.1.3 Composites

The composite is a substance produced by blending two or more materials together at the time of their manufacturing/production with a sole purpose of getting something better. The purpose can be reduction in unit weight of the final products —composite, and/or to improve their structural properties, such as strength, toughness, stiffness, and/or to make them fire and corrosion resistant, and/or to modify their conductance properties, and/or to enhance their durability characteristics, etc. The materials for the blending can be taken from a list of the natural or the man-made group. A few typical examples of the composites are concrete, plywood; fiberglass; reinforced materials, etc.

The concrete for instance consists of three different materials mixed together in the presence of water. The mix transforms into a solid mass after the hardening. The solid mass so formed possess properties very different from either of its basic constituent materials. The first major outcome of the process is that we can mold the material into any shape and size and after its hardening; it will be able to retain the desired shape. Additionally, we can use the concrete to bear any value of the applied load, whereas any of its constituent materials can never retain their assigned shape nor they are in a position to support the load as would be obtained from the final product. What happened inside the material that changed the way it behaves after the hardening? Any guess? The answer lies in the changed microstructure that takes place when water was added to the dry mix of cement, sand and coarse aggregates.

The cement in the dry mix, in the presence of water, binds the sand and the coarse aggregate particles together after its hardening. Some quantity of water added in the mix is consumed during the hydration reactions while the other part remains available to keep it workable, which permits its comfortable handling during the construction. The process initiates when water is added to the mix which triggers the hydration reactions in the cement. The products of hydration in the mix go on filling the empty spaces left in between the aggregates and, thereby, making the concrete denser and denser over a period of time; normally this process completes itself in the first 28 days. Figure 2.4a shows a typical polished section of a concrete sample that shows that how different constituents of concrete are packed together to form the solid mass. A closer look at the section reveals that it mainly consists of two phases: (a) aggregates of varying sizes, and (b) the binding fabric, which consists of an incoherent mass of the hydrated cement paste binding different aggregates together in the mix. So ideally, the hardened concrete mass should exhibit almost similar response under the different loading conditions, but it will not do so; neither it will be equal to the sum total of the load carrying capacity of all its constituents. Looks strange! This is the unique property that almost all composites will exhibit in contrast to the metals and other materials.

In concrete, for instance, we have three or more types of different materials, of different sizes and shapes placed randomly throughout the dry mix as well as in the hardened state. Each one possesses a different set of properties inherited from their parent materials. In the dry mix, all materials are lying defunct, as separate entities, without exerting any influence on each other. If we apply any load over this sample, it will support the load mainly by means of the direct contact that different particles and aggregates have with each other, along with the friction and their mutual mechanical interlocking. There remain many empty spaces in the form of voids in the mix. As the strength resides only on the solid part of the mix, the presence of voids often leads to a reduction in the load carrying capacity. However, the scenario changes entirely when the cement and aggregates are mixed in the presence of

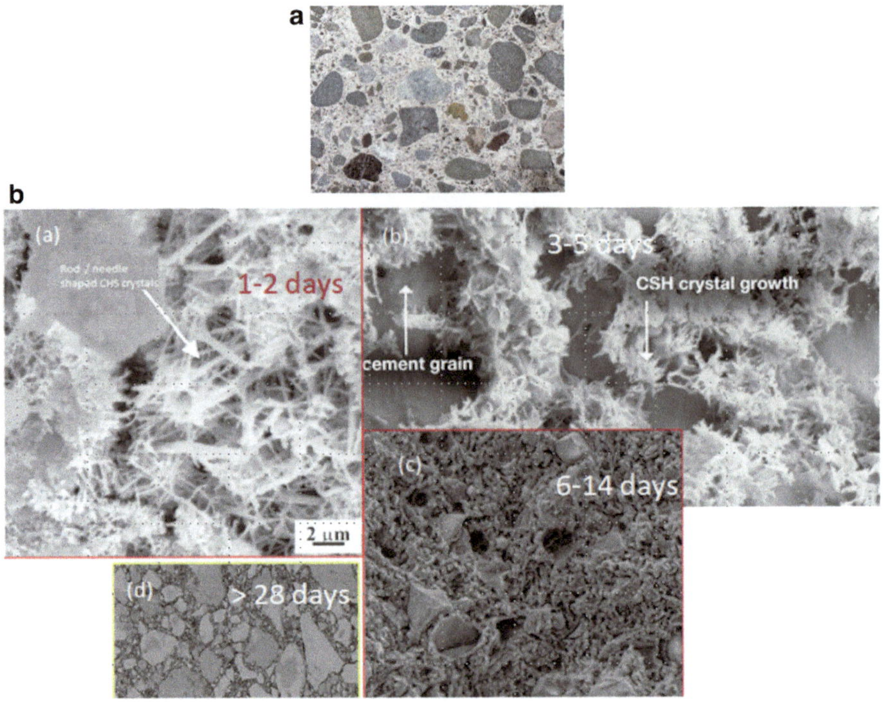

Fig. 2.4 **a** Typical polished concrete surface illustrating placement of coarse aggregates in the hydrated matrix of cement and fine aggregates (*Source* [12]). **b** Typical improvement that took place in concrete over a period of time: (a) Microscopic view after 1–2 days of its casting; (b) at 3–5–day period; (c) 6–14-day period and (d) after 28 days of hardening. In its hardened state, the concrete consists of hydrated cement paste, fine aggregates (usually, sand) and coarse aggregates (crushed stone particles). It formed when the cement hydrates in the presence of water and leads to the formation of different products of the hydration that, later on, fill in the empty spaces left between different constituents of the concrete to give it a unique solidness after 28 days. *Source* [12].

water. It leads to the formation of a very thin layer of separation around the surface of coarse particles in the concrete. This layer is called as an interfacial transition zone. Figure 2.5 depicts a typical microscopic view of such a zone developed around the aggregates. This zone typically extends to a distance of 10–50 μm at the surface of coarse particles in the concrete and is much weaker than either of the two main constituent materials of the concrete, namely the aggregates and the hydrated cement paste. However, it exercises a far greater influence on the mechanical behavior of concrete than is reflected by its size.

Actually, when water in sufficient quantity is added to the cement powder, it releases a lot of heat during its hydration. As a rough guide, it leads to a temperature rise of about 5.0–7.0 °C for every 50 kg of the cement quantity added in the mix. It will translate to a temperature rise of about 55 °C for a normal cement consumption of 400 kg/m^3 in a typical concrete mix used in the routine practice. Due to this heat,

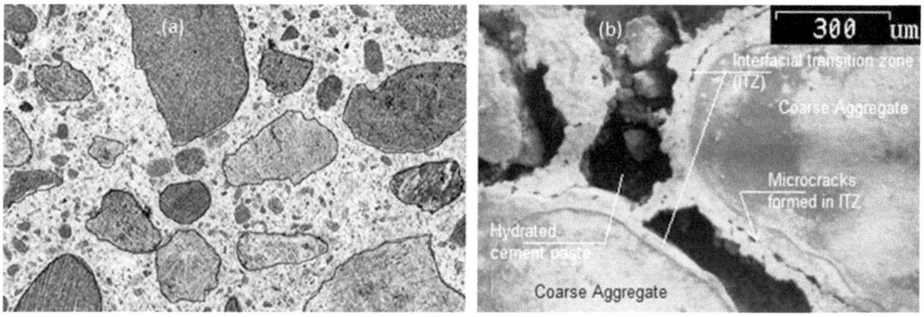

Fig. 2.5 Concrete microstructure: **a** Polished concrete surface depicting very thin lines of separation around coarse aggregates (shown in black lines); **b** microscopic view of ITZ around the surface of aggregates. This zone dictates the strength properties and the durability characteristics exhibited by concrete. *Source* [12]

the different physical properties of the constituent materials present in the mix lead to the formation of so called the region of discontinuity, especially, at the interface of coarse particles and the hydrated cement paste. It mainly occurs as the release of the heat in the concrete cause differential expansion of the aggregates embedded in the hydrated cement paste. Each particle in the mix undergoes different amount of thermal expansion depending upon their size and composition. And over a period of time, this expansion leads to the formation of a zone of separation around coarse particles when they return to the normal ambient temperature, as the cement paste by such a time already set to its final form.

Due to the formation of interfacial transition zone (ITZ) at the surface of coarse aggregates, they though stronger than both of the fine aggregates and the hydrated cement paste in the concrete usually have no direct influence on the concrete strength, except in the case where the concrete is made by using highly porous and weak aggregates, such as pumice. The large-sized coarse aggregates, particularly elongated and flaky ones, further add to the complications already initiated by the ITZ in the concrete, as all such particles have a higher tendency to accumulate water films on its surface. This phenomenon, known as bleeding, generally leads to the further weakening of the interfacial transition zone in the concrete, and it becomes comparatively more prone to the microcracking than other regions in the hardened concrete mass. This occurs because of a relatively high water–cement ratio that develops closer to the larger and/or flaky aggregates than away from it in the bulk of the cement paste. This phenomenon is found to be a major factor responsible for the shear-bond failure generally observed at the surface of coarse aggregates in the concrete when the concrete specimen is crushed in the compression.

At the microscopic scale, the chemistry of the concrete plays a greater role in controlling the behavior exhibited by the ITZ developed around coarse aggregates. The Portland cement in its anhydrous state consists mainly of angular particles typically in the size range from 1 to 50 μm. It is produced by pulverizing clinkers, which composed of a heterogeneous mix of several compounds formed by the high-temperature reactions between the calcium oxide and silica, aluminum oxide, and iron oxide. The clinkers on an average posses typically 45–60% tricalcium silicates (C3S), 15–30% Dicalcium silicates (C2S) along with 5–15% tricalcium aluminates (C3A) and some other compounds (C4AF) in range of 5–10%.

When the water is added to Portland cement, it rapidly starts dispersing calcium sulfates (C3S and C2S), and the other high-temperature compounds of calcium, such as C3A and C4AF into the solution. The Tricalcium silicate (C3S) rapidly reacts to release the calcium ions and the hydroxide ions into the solution. This chemical reaction is accompanied by the release of an enormous amount of the heat and it promptly produces the solution pH to over 12. This exothermic reaction is depicted in Eq. 2.1, which generally leads to the decomposition of C3S and C2S into the calcium trisulfoaluminate hydrates (C–S–H).

$$\begin{aligned}
&\text{Tricalcium silicate} + \text{Water} \Rightarrow \text{Calcium silicate hydrate} + \text{Calcium hydroxide} + \text{heat}\\
&2Ca_3SiO_5 + 7H_2O \Rightarrow 3CaO \cdot 2SiO_2 \cdot 4H_2O + 3Ca(OH)_2 + 173.6\,\text{kJ}\\
&\text{Dicalcium silicate} + \text{Water} \Rightarrow \text{Calcium silicate hydrate} + \text{Calcium hydroxide} + \text{heat}\\
&2Ca_2SiO_4 + 5H_2O \Rightarrow 3CaO \cdot 2SiO_2 \cdot 4H_2O + Ca(OH)_2 + 58.6\,\text{kJ}
\end{aligned} \tag{2.1}$$

Within no time, the needle-shaped crystals of (C–S–H), called as ettringite, start forming in the cement paste, which over a period of time, start filling the empty spaces existing in the matrix, formerly occupied by water and dissolving cement particles. Afterward, depending on the alumina-to-sulfate ratio used in the production of the Portland cement, the ettringite may become unstable and breaks up to form the monosulfoaluminate hydrate. This ratio plays an important role in controlling the quantity of monosulfoaluminate hydrate being librated/formed in the concrete mass. The van der Waals forces of attraction existing between various solid products formed during the hydration process contribute to the strength exhibited by concrete. The products of the hydration generated in the cement paste tend to adhere strongly not only to each other, but likewise to the other solids having a large surface area, such as calcium hydroxide, anhydrous clinker grains and aggregates, which finally build a microstructure of the concrete.

The prevailing high water–cement ratio near the surface of the coarse aggregates often leads to the formation of relatively large-sized crystals during the hydration process. The relatively less adhesion capacity of these large-sized crystals in the already fragile ITZ further lowers the concrete strength because of their smaller surface area and consequential weaker van der Waals forces. These crystals also act as a preferred cleavage sites for the failure owing to their inclination to form an oriented structure. These factors ultimately lead to further weakening of already fragile interfacial zone at early ages in the concrete. However, as the hydration process progress over the time, the poorly crystalline C–S–H and second generation of smaller size crystals of the ettringite and the calcium hydroxide generated by the slow chemical reactions in the cement start filling the empty space already existing between the frameworks created by the larger ettringite and the calcium hydroxide crystals. This leads to densification of the concrete microstructure over a period of time (see, Fig. 2.4b). Such interactions are the prime source of the strength gain by the concrete with time. It is interesting to note that out of various compounds released during the hydration process, only the calcium silicates contribute to the strength of the concrete—C3S is responsible for most of the early age strength of the concrete, while C2S contributes toward the strength at the longer durations, such as 28 day or more. Think, why?

Unlike the microstructure of concrete, if we look at other natural materials such as stones, etc., these usually do not have any interfacial transition zone in the matrix rather they are composed of a large number of grains just similar to the metals, though their sizes are much larger and coarser than that exist in the case of the metals. These grains are mostly held together by means of the secondary bonding, but due to their larger grains, the stones will have relatively weaker internal bonding than the metals. The presence of voids, cavities, mineral cleavages and other microfractures that developed during their formation also becomes responsible for the further reduction in the strength that they would exhibit otherwise. These lines of discontinuities also control the direction along which the stones fail under the action of applied load. Normally, the fine-grained varieties of rocks/stones are stronger and more durable than the rocks/stones which are coarse grained in their texture. Think, why?

2.2 Material Behavior Under Different Stress Conditions

Materials being useable part of the matter inherit almost everything from their parental source of extraction. Their chemical/mineralogical composition plays an important role in deciding the type of behavior they would exhibit. The processes used in their production also have a significant influence on their behavior. The most important aspect that these parameters will control is the microstructure of a material. There exists a coupling between the mechanical properties of the material and the microstructure it possesses. Whatever modification done in the one parameter has a significant effect on the other. This phenomenon allows the material technologist and engineers to modify the material properties by making suitable alterations in the microstructure, which in general indicates the type, extent, size, form and distribution of different phases within the material matrix. As the strength of the material resides in its solid part, the voids therefore left in the material because of any reasons are detrimental to the strength. The only way out therefore is to modify the material microstructure by filling these voids, etc. by adding some other material particles at the time of their production that are finer than the main constituent units/grains of the material. If we are able to modify the microstructure and make the material more dense, some significant changes in the material properties can be made. The following sections describe different types of the responses that materials exhibit depending upon their microstructure.

2.2.1 Tension

Metals: Microstructure of the metals varies from the fine textured to coarse one. It contains a large number of mutually packed grains of various sizes and shapes, held together by means of the metallic bonding between atoms constituting such grains. The size of a grain can range from about 1 μm (small grains) to 100 μm (large grains). As the strength of any metal is a measure of the difficulty to move the dislocations through and between the grains, small-sized grains indicate a higher metal strength as the chances of occurrence of internal voids, cavities and other dislocations will be very small in such a setup. In case, the size of grains is smaller than 1 μm; the size of dislocations begins to approach almost the size of the grains itself; thereby giving almost nil possibility of any dislocation formation. In such cases, there exists a pure crystalline structure in one or other form in the metals and their strength approaches the maximum theoretical values computed on the basis of bonding characteristics.

The metal tensile strength is indicated by a force required to move the atoms lying along the slip line away from each other; it will be higher if we need more effort to move the atoms. It is worthwhile to recall that dislocations are simply an irregularity within a crystal structure of metals, which contain an abrupt change in the arrangement of atoms. The movement of dislocations allows atoms to slide over each other at low stress levels and is known as a slip. And, the line along which it happens is called as slip line (see, Fig. 1.11). Given enough stress and thermal energy, dislocations will move easily throughout the crystalline grains, resulting in permanent distortion of the grain itself. However, once a dislocation reaches a grain boundary, it has nowhere to go. The grain boundaries become a barrier to further movement of dislocations. Thus, an easy way to improve the strength of a material is to make the grains as small as possible, which increases the number of grain boundaries within the system. Since the lattice structure of adjacent grains differs in orientation, it requires more energy for a dislocation to change directions while passing from one grain to another. Moreover, smaller grains also contribute indirectly by providing relatively higher ratios of surface area to their volume, which means a greater ratio of the grain boundary to the dislocations. It also reduces the chances of the possible pile-up at the boundary, thereby increasing the amount of applied stress necessary to move dislocations across a grain boundary. Impeding this dislocation movement will hinder the onset of plasticity and leads to increase in the yield strength. There exists an inverse relationship between the grain size and the yield strength of metals.

Stones: Like the metals, stones also composed of many grains of various sizes and shapes. But the stones are not as strong in tension as the metals are. The principle reason for this is their coarser grain size in comparison to the metals where these are comparatively very fine. Moreover, these are formed from various chemical compounds—not the metallic bonding at the atomic level as exists in case of the metals—generally, crystallized in the form of a tetrahedra with the silicon oxide in their crystal lattice. Secondary bonding is the prime force which holds various stone grains together. As this type of the bonding is weaker than the metallic bonding, stones exhibit comparatively lower tensile strength. The presence of other defects, such as microcracks, joints, veins, fissures, dikes, etc. causes further degradation of their strength properties. Unlike the metals, they may act stronger along one of their axis while the strength along some other axis may be almost negligible. Due to these reasons, the stones from the igneous rock are usually stronger than those obtained from the metamorphic or the sedimentary rocks.

Polymers: Unlike both the metals and the rocky materials, polymers have entirely a different form of microstructure. These contain a large number of long polymeric chains cross-linked and tangled among each other. Depending upon the extent of their cross-linking, they exhibit a wide variety of properties. However, the absence of free electrons in the material imparts them zero conductance to both the heat and the electric current. Normally, they possess a high unidirectional tensile strength along the direction of the polymeric chains owing to the presence of a strong covalent bonding among their different units. The strength along the other direction is highly influenced by the degree of orientation of the molecules and their cross-linking, caused by the covalent bonding and the secondary bonding—mainly the van der Waals forces among different elements of the chains. Depending upon the extent of the covalent bonding, their stiffness can be as high as 1000 GPa (as found in case of diamond) and it can be as low as 1 GPa (for the simple hydrocarbon based materials, like paraffin). It is around 3–4 GPa for some types of the polymers (for instance, plexiglass) which do not have any cross-linking, whereas it is around 8–12 GPa for the cross-linked polymers. However, for getting the desirable mechanical properties, we need to have a minimum of 500 monomers in any given polymeric chain. The number of such monomers in any chain defines their polymerization, which usually ranges from 10^3 to 10^5 for the normal practical applications. The wood and bamboo—all natural materials made up from the biopolymers, for instance, are very strong in tension in the direction of their fibers than the axis normal to their own.

Composites: The response of the composites, unlike all the materials described above, is very sensitive to the microstructure and highly dependent on the material composition used to produce such composites. For instance, the plywood is a composite made by pasting and pressing a number of thin wooden sheets—called as veneer together in the form of boards. As the wood has a tensile strength only along its fiber axis, pasting the veneers normal to each other, alternatively, will help to get plywood which will have a comparable tensile strength along its length and width.

Concrete being the very common building material will be taken here to illustrate the role of various constituent materials used to produce it and their effect on the microstructure and subsequently, on their mechanical behavior. As described in the previous section, a concrete sample always has a number of internal microcracks, especially over the surface of coarse aggregates even when there is no external load acting over it. This type of cracking creates a very fine interfacial transition zone around the surface of coarse aggregates. This zone extending up to a distance of 30 microns around the surface of coarse aggregates possesses more capillary voids in the network of large-sized calcium hydroxide crystals than in the bulk of the concrete matrix; it is more porous than any other region in the concrete mass (See, Figs. 2.4b and 2.5). The exception may be a lightweight concrete which contains pumice as coarse aggregates that is too porous to enable us to reduce the unit weight of concrete from 24 kN/m^3 to as low as 14–18 kN/m^3. This unique feature of the interfacial transition zone (ITZ) plays a crucial role in dictating the constitutive relationship of a concrete specimen, especially its tensile strength. Nevertheless, the improvement in the density of the ITZ over a longer duration, especially after one year or so, helps to improve the concrete strength in general and its elastic modulus in particular. The slow chemical reactions going on within the concrete microstructure contribute considerably toward the strength gain over the longer times.

Concrete is very weak in tension and fails abruptly. It can again be attributed to the presence of a very porous ITZ around coarse aggregates, which acts as a cleavage plane for the microcracks in the zone to develop into cracks more easily under the tensile stresses than the compressive stresses. Under a tensile load, the hydrated cement matrix in concrete tries to move away from the aggregates. The porous and fragile zone easily allows it to happen due to the weaker forces that holds various constituent units of concrete together. The consequence is that concrete is brittle material under a tension load which leads to a sudden failure once the tensile strain in the specimen reaches the material threshold limits. A typical response of concrete and its comparison with stones,, etc. is depicted in Fig. 2.6.

It is important to note that the size of coarse aggregates and its grading; cement–content and water–cement ratio; humidity and curing conditions of concrete have a considerable influence on the extent of the microcracking developed in the concrete. A concrete mix containing poorly graded aggregates is more prone to the segregation. Similarly, the concrete containing a larger fraction of flaky and large-sized coarse aggregates tends to have a thicker water film around them, especially beneath the particle, even under an identical set of casting conditions. The water–cement ratio used in the production of concrete, likewise has a significant effect on the size of crystals that would develop in the matrix. An increase in the water–cement ratio leads to the formation of coarser crystals in the concrete. The ITZ developed under such conditions is relatively more porous and weak in nature and it acts as a preferential cleavage plain for the microcracks in the zone to develop into cracks more easily under the tensile stresses. So, any technique that helps to produce concrete with a low water–cement ratio tends to have smaller size crystals of the hydration products; thereby giving an enhanced concrete strength, owing to the presence of relatively higher van der Waals forces acting in the matrix.

2.2.2 *Compression*

A compression force tends to bring the two or more particles closer over which it is applied. Various constituent units of a material, therefore, come nearer to each other if some force of compressive nature is applied over them. Once that starts

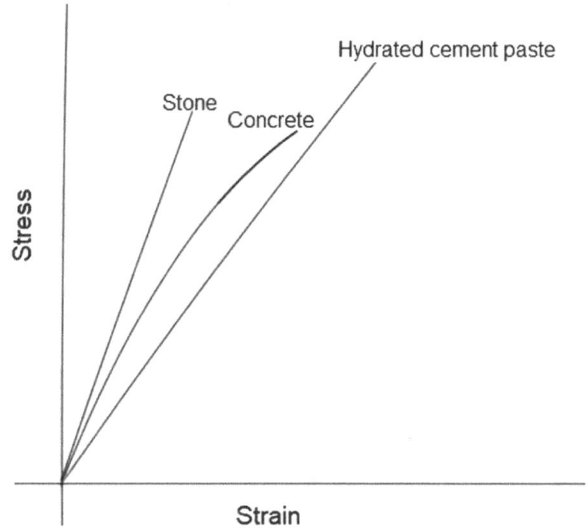

Fig. 2.6 Typical tension stress–strain response of different materials—the stones and the hydrated cement paste exhibit a linear-elastic response upon stressing, whereas the composite/concrete does not. The presence of the interfacial transition zone around the surfaces of coarse particles in the concrete is the prime reason for this change in the material response, when it is strained beyond the elastic range in comparison to the stones and the hydrated cement paste. The interfacial transition zone does not exist in the case of the stones and the hydrated cement paste. *Source* Author

happening, it naturally leads to a better bonding and increased direct contact between them, which further leads to an enhanced frictional interactions among them; especially so, if the constituent units are discrete particles, e.g., coarse aggregates, sand, soil particles, concrete, etc.

Irrespective of the type of the force, whether tensile or compression, the volume of the material specimen remains same when it deforms under the action of the force. Under a compression force, its length will reduce so it has to bulge sidewise in order to keep its volume constant. It automatically leads to the development of tensile forces, normal to the loading axis, when the constituent units of the material start moving away from each other during the bulging process. In case the force is acting vertically on the specimen (along its axis), it produces the tensile force along its horizontal axis. This phenomenon is called as Poisson's effect, which have a major role in controlling the material response. The brittle materials fail abruptly as they usually are weak in tension, which also put an upper limit to their compression load carrying capacity. The coarse grains in the case of the brittle materials are the main culprit for their lower tensile capacity.

Metals on the other hand have a finer grain structure in comparison to the stones and the composites. As a result, they are strong in tension. This also led to their high compression strength. Just like their load–displacement response in tension, they have a linear relationship up to the limit of proportionality and behave as an elastic material, followed by a nonlinear response till their rupture. Up to a certain level of the tensile stress, the grain structure in the metals remains intact and undisturbed. They transfer the stresses safely grain- to- grain and respond elastically as the internal forces operating at the atomic level bring various constituting units back to their original level once the external forces are removed. However, when the stress level is increased beyond a certain limit, the internals crystal defects, such as dislocations, etc. start affecting their tension response. Usually, the metals enter their nonlinear (plastic) range where they start deforming at a very large rate in comparison to the rate of loading applied to the specimen. In this range, the dislocations start moving fast and end up at the grain boundaries. The piling of dislocations at the boundaries results in the plastic deformations. Almost similar trends are exhibited by the metals in their compression behavior; it is a linear-elastic up to some stress level and if the specimen has a enough cross-sectional area to prevent crushing, it will enter the nonlinear plastic response, followed by failure when their internal microstructure fails to transfer the stresses safely (Think, how does the Poisson's ratio play its role here?).

Stones and polymers also exhibit a response almost similar to the metals, albeit with a lower load carrying capacity and without any significant nonlinear plastic range. They usually behave as a brittle material. Their coarse grains and heterogeneous chemical composition play an important role to make them brittle. Such types of materials usually have a relatively lower tensile capacity, which also indirectly put a limit on their compression capacity. In all practical applications, however, an appropriate selection of the cross-section size and shape of structural members usually serve our purpose to transfer the compressive load safely through them. Their member sizes are selected so as to limit the stress produced by external loading to that a safe value permitted by the material (called as their design strength).

Composites exhibit entirely a different response depending upon the constituent materials used to produce them. They can act as brittle materials as well as ductile ones. Because the brittle materials fail abruptly without showing any sign of distress —a sort of warning of impending failure—the composites are never designed to perform in such an undesirable fashion. They can be proportioned and designed to behave as ductile materials. Consider the case of concrete, the most commonly used building material due to its versatile nature. Recall the concrete microstructure described in the previous section and checks how the presence of aggregates in the cement hydrated matrix affects the tensile response of a concrete specimen. The presence of interfacial transition zone (ITZ) over coarse aggregates in the hardened matrix leads to a reduction in the load carrying capacity of concrete in tension and nonlinear response in compression, with its tensile capacity as one-tenth of the compression strength. On the other hand, if we test the hydrated cement paste and the aggregates in isolation, they would exhibit an entirely different response—linear-elastic, followed by a sudden failure. These two materials also have a higher stiffness in comparison to the concrete. This happens because of the presence of the interfacial transition zone in the concrete, which the aggregates and the cement paste do not have. This zone acts as a bridge between the hydrated cement paste and the coarse aggregate in the hardened concrete and is an only medium to transfer stresses across its different phases. The presence of the voids and other microcracks in the ITZ do not allow an effective stress transfer across its hollow space. It therefore leads to the diminution in the load carrying capacity of the concrete, irrespective of the load type. It also influences the elastic modulus of concrete. Considerable energy is needed for the formation of new microcracks and extension of those already existing in the ITZ of the concrete mass under a compressive load than the tensile loading where these cracks propagate rapidly and that happens at a much lower stress level in comparison to the former type of the stress.

If we observe the compression load–displacement response of a concrete specimen, it reveals a linear-elastic behavior of concrete when it is stressed to one-third of its ultimate strength. Generally, the microcracks in the ITZ remain undisturbed up to this stress level and behave elastically—they open up and regain original position on removal of the load. These cracks start growing in length and size, if the stress level is increased further; the prevailing stress concentration at the crack tips and other large voids in the specimen makes the things easier for them to respond this way. These cracks, however, confine themselves to the ITZ, and did not run into the mortar matrix. To this point, the crack propagation in the zone remains more or less stable under the prevailing stress conditions in the concrete specimen.

When the stress level in the specimen is gradually raised to 50% of its ultimate compressive strength, the shape of the compression response curve starts deviating from the linear-elastic response exhibited earlier. This deviation can be attributed to the extension of the existing cracks in the zone to the adjoining mortar matrix as the load is increased gradually. And, when it reaches about 75% of the ultimate compressive strength, the available internal energy in the system exceeds the crack-release energy required to initiate the cracking. This process results in a rapid growth of the cracks, which finally spread into other regions of the matrix until these starts joining the cracks originating from the ITZ of surrounding aggregates and subsequently leads to the unstable propagation of the cracks in the adjoining areas of the concrete specimen. The stress level corresponding to this stage is called as critical stress. At this stress level, the concrete possesses a highest ever volumetric strain and a slight increase in the magnitude of the stress beyond this point result in a reversal of the direction of the volumetric change. It is accompanied by a sudden volumetric expansion of the concrete specimens, followed by its crushing as the load reaches the ultimate compression strength. A typical compression stress–strain response of concrete specimen is depicted in Fig. 2.7.

The value of the crushing strain remains more or less same for all types of concrete, whether a normal or the high strength concrete. Normally, the concrete ruptures abruptly when it is strained beyond its limited tensile strain, generally ranging from 0.00012 to 0.00016. The failure, however, is more gradual when it is compressed. Normally, the concrete specimen fails to support any additional compression load when it is strained beyond a strain value of 0.0035. This occurs irrespective of the concrete grade. The shape of the stress–strain curve becomes steeper around a strain value of 0.002 as the strength of concrete is improved, for instance from M25 to M30 and so on, and thereafter, the concrete exhibits a strain-softening response (see, Fig. 2.7).

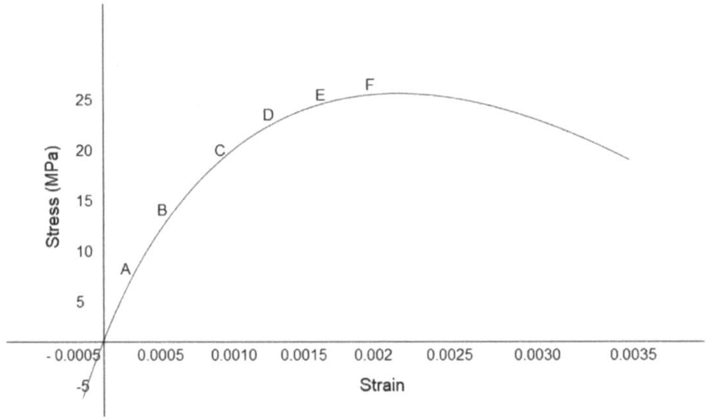

Fig. 2.7 Concrete response at different compression stress levels—the figure schematically shows a stress–strain response exhibited by concrete at various stages of uniaxial stress conditions. When the concrete specimen is compressed to about 30% of its ultimate strength (stage A), the response is linear-elastic; followed by a slight deviation in the response when the stress level is maintained in a range of 30–50% of the ultimate concrete compressive strength (stages A to B). At a stress level of 75% of the ultimate compressive strength (stage C); the concrete shows a significant decline in the slope of the stress–strain curve and it possesses a higher value of the volumetric strain at this stage. Thereafter (stages D to F), it exhibits a sharp change in the slope of the curve and it generally fails to support any further loading. It is accompanied by the strain-softening after reaching the stage F. *Source* Author

It is worth to note that the improvements made in the microstructure of concrete by addition of mineral admixtures, such as slicafume, rice husk ash, fly ash, etc. at the time of production make the concrete more homogeneous in texture. These particles being finer than the ITZ occupy the free space and also being pozzolanic, starts forming new compounds that help to reduce the thickness of ITZ over a period of time. Because of an improved microstructure, the concrete produced using such admixtures, for instance the high strength concrete, show a steeper stress–strain response, comparatively less volumetric dilation and most importantly, reduced permeability than the normal concretes. The recent experimental results revealed that by making the necessary improvements in the microstructure, it is possible to get the concrete compression strength as high as possessed by the steel; however, its tensile strength remains more or less 5–10% of the compression strength.

2.2.3 *Shear and Bending*

Shear is a force that acts at any point in the beam normal to its member-axis. It always acts in pair—one on each side of the point under consideration, but in the opposite directions so as to satisfy the equilibrium conditions—and tends to slide the two segments of the beam vertically up and down across that point. The rotational effect of this pair of the forces is known as bending moment. Shear and bending always coexist in the beams. Unlike the tensile strength and the compression strength, shear capacity and bending strength of material are not pure material properties; beam dimensions, its span, loading, support conditions and lateral support conditions of the member being tested also play an important role in controlling their values. These two values, therefore, depend more on the structural parameters of the member being tested than the material properties used in its production. Both the beam shear capacity and the bending strength at the end of the story depends on the tensile strength of material being used to fabricate or make such beams as the tensile strength of any material used in civil engineering applications being the lowest among all its strength values normally put a cap on the load carrying capacity of the member, be it a shear or the bending strength. Therefore, inherently both of these parameters carry brittleness with them and lead to an abrupt failure in case the stress caused by the external load reaches its permitted value.

It is, however, possible to increase the strength by suitably reinforcing the material by means of some internal or external reinforcement, such ass steel bar, steel fibers, stiffeners. This type of approach is used commonly in practice, to increase the shear and bending strength of the beams fabricated either in mild steel or produced by using the concrete. The final product basically acts as a composite with all the constituent components acting in unison to achieve the main objective of the load transfer through the member made up of such materials.

As the concrete is inherently weak in tension, it will not be able to support any load that produces tensile stresses in the member. Any member if loaded normal to its axis (say, by gravity loading in case of the beams) always develops tensile stresses below its neutral-axis along with a matching compressive stresses above the axis. Once the internal energy stored in the member because of the external loading exceeds the crack-release energy of the concrete, cracks start developing rapidly in the concrete. The prevailing tensile strain conditions below the neutral-axis of the member further exaggerate the problem. Therefore, it will fail abruptly once these stresses reach its limiting tensile capacity of the material. In all such cases, it becomes mandatory to use some material having a high tensile strength, like steel, etc. in the tensile zone of the beam to meet the moment demand. This type of reinforcement provided in the member is called as rebars. We can also use short, discrete steel fibers mixed in concrete at the time of its production in place of rebars to meet this purpose. Reinforcing concrete this way helps to transform the overall flexural response of the concrete section from a brittle to ductile one, all depends upon the percentage of the steel provided in the section or dosage of the steel fibers used in the concrete production.

The rebars provided in the tensile zone of the concrete member take up the tensile forces through various bond mechanisms, viz: chemical adhesion, frictional resistance, mechanical interlock, etc. and subsequently, these rebars prevent the widening, and extension of the cracks toward the neutral-axis of the member under increasing load. If the bond between the rebars and the concrete became inadequate because of any reasons; it results in slippage of the rebars and often leads to redistribution of the stresses in the member. It is generally accompanied by the crack formation in the tensile zone of the member. The mathematical models for proportioning RC beams have been developed over the past 150 years and are well documented in the published literature [12].

2.3 Parameters Affecting the Material Strength

Material strength as pointed out in previous sections depends entirely on its microstructure. Therefore, all parameters that influence microstructure of a material will affect its strength. Unlike the density of a material, its strength is an intensive property; it did not depend upon the quantity of material being tested. However, for some materials, it is a function of many other independent variables. The metal strength, for instance, gets affected by the change in the temperature; normally it reduces when the metals are heated beyond some value or cooled below some temperature value. Similarly, the size and shape of a specimen; the rate and duration of the loading and for some materials, the direction along which these are being tested affects the material strength. In order to get a design value of the material strength, therefore, standardization of test setups and related procedure is of utmost importance. This section describes different parameters that affect the material strength.

2.3.1 Size and Shape of Specimen

The cross-sectional dimensions of the test specimen have a pronounced effect on its strength. In fact, it increases with the increase of the member cross-sectional area (loading face of the specimens), which enables them to support higher loads. As pointed out in the previous section, the tensile strength of a material is nothing but a sum total of all such chemical bonds that connects one cross section in the specimen with the one adjacent to it. An increase in the cross-section area means that more numbers of such bonds are available that would keep the material intact. So, there exists a linear relation between the cross-section area of a member and the load it would support during the testing. However, as the length or height of the test specimen is increased, it starts exhibiting reduction in the compression strength. It is worth to recall, that the tensile and compression strength of any material are related to each other. Whenever any specimen is loaded in compression, tensile stresses always produce on a plane normal to its loading axis, which indirectly set an upper cap on the compression strength.

When any specimen is tested in compression, the state of stress has to be uniform over its full cross-section to get a true value of the strength, but for the short specimens (having a low height-to-width ratio) it is very difficult to obtain such uniformity. As the specimen is loaded axially, frictional forces start developing between its top and bottom faces and adjacent steel plates of the testing machine, which restrains the free movement of the specimen in the lateral directions. This effect is known as *platen-restraint* and it leads to increase in the compression strength as the radial shear force being induced in the specimen during the testing are additive to the tensile stresses acting there because of the Poisson's effect. However, this beneficial effect goes on reducing as the height-to-width ratio of the specimen is increased, especially near its mid height where the sample will fail earlier than otherwise for a short specimen.

2.3.2 Type of the Loading

Microstructure of material responds differently to different type of loading. Any loading that brings the basic building block of the material closer will improve the mechanical properties. A tensile force tends to move the points away from each other, so it reduces the electrostatic forces acting at the atomic level in materials which finally manifest in the form of their reduced tensile capacity. The movements of dislocations in the lattice further exacerbate the problem. On the other hand, if the loading confines the lattice or brings the particles responsible for holding the lattice together, closer to each other will improve the load carrying capacity of the material. Compressive forces applied to materials bring into being this type of effects and they exhibit increased capacity; however, in the absence of any confining effect, their limited tensile capacity sets a limit on the maximum possible achievable compressive strength, as happened in the case of the concrete.

2.3.3 Duration of Loading

The duration and the rate of loading, whether instant or sustained, have a significant effect of the strength parameters. Whenever any specimen is loaded at a faster pace (simulating impact type conditions), they usually exhibit a higher elastic modulus and compressive strength in comparison to the case when they are loaded gradually over the long durations, say one year or so (i.e., sustained loading conditions). When the load is applied gradually over longer durations, it also improves the failure strain of the material as well as its post-peak response curve, which changes to a flat curve from the strain-softening response exhibited by it in the high strain rate loading.

2.3.4 Stress Conditions

Changing the stress conditions from uniaxial to multi-axial improves the mechanical response of materials. The confinement provided by the multi-axial stress conditions provides stability to the material microstructure by preventing any free movements that could occur otherwise in the case of the uniaxial or the biaxial stressing. For instance, the biaxial stress conditions lead to a 25% improvement in the concrete compressive strength in comparison to the case when the similar specimen is uniaxially stressed. So, due care should be taken while deciding the loading conditions during the test operations.

2.4 Material Constitutive Models

Constitutive modeling is the mathematical description of how materials, respond to loadings. It forms a backbone of any engineering analysis and design problems. In this process, various mathematical, analytical and/or statistical methods and techniques are used to model the behavior of a material under different loading conditions, which is then used to predict the response and the load carrying capacity of structural members. However, it is very difficult to model exactly and precisely the way materials responds. Normally, an attempt is made to model the material response in a way that it should be sufficiently accurate, yet not unnecessarily complex and computationally expensive. Additionally, it should be simple to use and apply to a variety of field/design conditions. So, while developing any material model, following questions can help the analyst to arrive at a suitable model:

- Is the model relevant for describing the physical phenomena being examined?
- Does the model produce sufficiently accurate predictions for the given purpose?
- Is it possible to devise and implement a robust numerical algorithm based on the model?

It the answer is 'yes' to all the questions, then most likely, it is going to serve our purpose and we can get sufficiently accurate results from the model. For example, it is appropriate to claim that the behavior of concrete can be represented by an elastic model, but it does not make sense to claim that it will act as a pure elastic material. It can also be modeled as elasto-plastic material. Both models will be good enough to describe the way concrete will respond during the loading process. It all depends upon the purpose and the required precision of the final prediction model and its type of applications. The first one is routinely used in the design of concrete structures where purpose is to retain liquids, such as water storage tanks. Whereas, the elasto-plastic concrete models find their use in the limit state based design procedures used to proportion flexural members, axially loaded members, etc.

A constitutive model is supposed to satisfy some of the fundamental conditions to be a truly representative of the material response; namely, **(1)** Principle of coordinate invariance—Constitutive relations, as well as other relations between different physical entities, should not be affected by arbitrary coordinate transformations; **(2)** Principle of determinism—The stress field in any given member is determined entirely by the history of the motion of the member as established through some set of experimental observations or a set of the result data from simulation studies; **(3)** Principle of material objectivity—Constitutive relations must not be affected by arbitrary rigid body displacements that is superposed on the actual motion of the member; **(4)** Constraints of material symmetry—Response functions should remain unaffected by certain rotations of the chosen reference configuration arising from the material symmetry; for instance, as in case of material isotropy; **(5)** Dissipation inequality—Dissipation of energy in the system should never be a negative quantity. While developing any material model, an attempt should be made to meet as many listed conditions as possible. If all the five conditions are met, then most probably the model will be able to predict the solution near to the true value and it will be applicable for most of the field constraints.

We can predict the response of any structural member easily and only if constitutive model of different materials which constitute the member is known. The equilibrium and compatibility of displacement conditions are then applied to capture the member response and/or its load carrying capacity. Fundamentally, an analyst needs to know first how the material will respond to tension and compression stresses. Once that is known, these can be used to derive the material response under some other types of stress conditions, e.g., shear and bending. A set of the generalized material models under the two different fundamental stress conditions are presented in the following sections:

2.4.1 Tension Models

All materials perform differently under the tensile loading, very unique to each and every type depending upon their chemical composition and microstructure. Some fail suddenly (e.g., concrete) without giving any sort of distress signal, while the others (e.g., mild steel) goes on elongating till their rupture at the end; thereby giving ample time to the user to respond and take some corrective actions to mitigate the ill-effects of applied load. Broadly, following types of models can be categorized to represent the material response to a tensile force.

2.4.1.1 Multi-linear Relationship

This type of a load–displacement response curve is very unique to the ductile materials; these can be metals, polymers or specially designed composites, such as reinforced concrete, steel fiber-reinforced concrete, etc. Usually, they initially exhibit a linear-elastic response up to a certain limit (ε_1); then, enter into the plastic range and go on elongating till the rupture. A typical such type of response is depicted in Fig. 2.8.

The slope of a tangent drawn at the origin of the curve gives the stiffness of the material. The change in the slope of the curve at different strain or displacement values indicates a gradual switching of the load transfer modes from one to another, especially in case of the composites, e.g., the reinforced concrete and granular

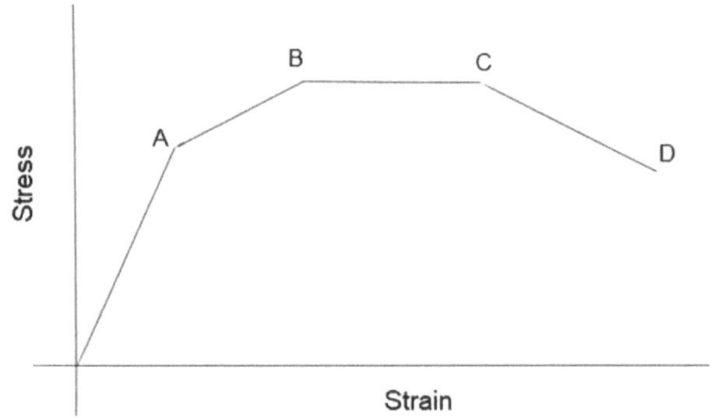

Fig. 2.8 *Typical multi-linear model*—it consists of a set of idealized straight lines joining each other at the characteristic key points (A, B, C, and D), depicting the real response of the material being examined. *Source* Author

materials, e.g., soils. For the metals; however, it may be a change in the position of atoms during the initial linear straight curve; followed by movement of dislocations in the grains of a lattice structure and then, they pile-up at the grain boundaries and finally, the rupture when their lattice structure is no more able to accommodate such large displacements. It is worth to note that some material may undergo a strain-softening or strain-hardening depending upon their microstructure. Normally, the slope of the response curve post-yield indicates the type of material behavior—strain-softening or strain-hardening. If the slope shows a rising trend, it indicates the strain-hardening; otherwise it would be a strain-softening behavior.

2.4.1.2 Bilinear Relationship

This is a simplified version of the multi-linear material response model and contains only two segments—one, representing the linear-elastic behavior at low stress levels and two, a post-peak curve representing the material response to the loading till its failure. Some polymeric materials; inadequately reinforced materials, such as SFRC with low fiber dosage, RC beams having a low percentage of tension steel in the section, etc. respond this way. Figure 2.9 depicts a typical bilinear response exhibited by some of the materials. The slope of the second line segment of the response curve indicates the material strain-hardening or softening characteristics depending upon its slope. It will be softening when the load carrying capacity of the material reduces with increasing strain. The negative post-peak slope of the curve is indicative of a strain-softening behavior. Normally, such types of materials are less tough than those who exhibit the strain-hardening. It needs extra effort and cost to make materials behave in the strain-hardening mode and even to get strain-softening in the case of the brittle materials.

2.4.1.3 Linear Relationship

Unlike the previous two models, the linear models are very unique to the brittle materials which fail to carry any additional load on reaching their limiting strains.

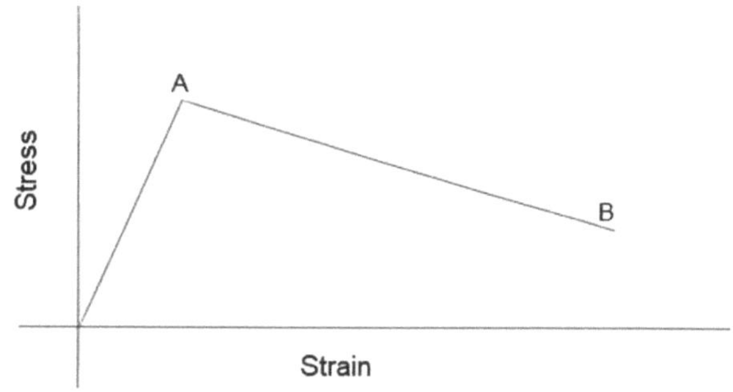

Fig. 2.9 *Typical bi-linear model*—it consists of two idealized straight lines joining each other at characteristics key point (such as A), depicting the initial elastic range only, and followed by the plastic response till its failure at B. *Source* Author

They abruptly stop to carry the load increments and breaks into two or more segments. As such, there will be no post-yield curve for such type of materials. Some typical examples are stones, bricks, and the plain concrete. A typical stress–strain response for such type of materials will have only one straight line ending at the strain ε_1; if we ignore the second line starting the post-yield point in Fig. 2.9, it will be idealization of the material response using this model. Usually, they undergo a linear-elastic response up to certain limit and then, fail. There is no or almost zero post-peak response. The slope of a tangent drawn at the origin of the curve gives the stiffness of the material. As this type of the response curve has the lowest area, the materials that are modeled using the linear models possess smallest toughness. The limiting strain for most of the brittle materials varies from 0.00012 to 0.00016.

2.4.1.4 Drop-Constant Relationship

It is a special type of the model applicable for the materials which on reaching certain strain levels suddenly losses their peak loads and post-peak, they posses almost a constant value of their residual carrying capacity. This line can be an average of the sloping post-peak curve drawn at the point corresponding to the material residual load carrying capacity. Figure 2.10 depicts a typical response for such type of materials. There are some composite materials which are inherently brittle in nature and fail to carry any additional load on reaching their limiting strains. But after reaching this strain level, they continue to support the load, albeit with a smaller load carrying capacity may be by means of the dowel action or the internal reinforcement present in the material. Steel fiber-reinforced concrete with low fiber dosage is a typical example of the materials that would exhibit this type of response.

2.4.2 Compression Models

Materials response is entirely different under a compression force in comparison to the tensile force. They can fail following a brittle mode of the failure, while some

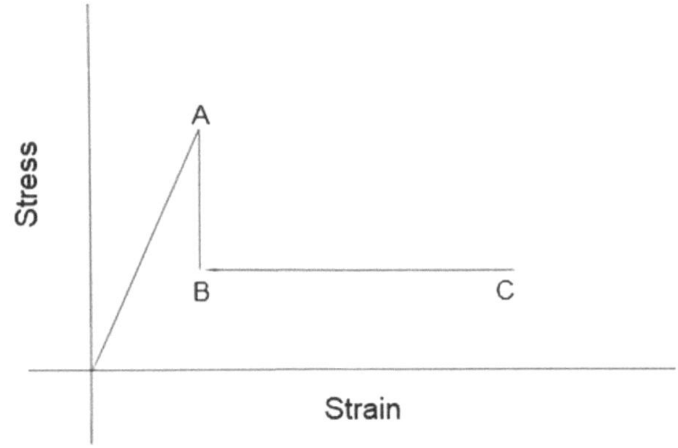

Fig. 2.10 *Drop-constant model*—it consists two or more straight lines describing the initial elastic response of the material which loses its strength post-yield, but continues to support the load by virtue of other stress-transfer mechanisms being activated at higher strain levels. *Source* Author

may exhibit a ductile mode. The materials which are inherently weak in tension will have a limited peak load in the compression and they generally soften in their post-peak range. And, in the case of a good tensile strength or a residual tensile strength, they may exhibit strain-softening or may be strain-hardening.

Plain concrete is a good example of the materials where the peak compression load carrying capacity of the sample is greatly limited by their poor tensile strength and if some quantity of short steel fibers is added in the concrete at the time of its production, an improvement in the tensile capacity of the material also led to an improved post-peak behavior of the sample (see, Fig. 2.10). The resultant toughness of the material can increase by a factor of two to three in comparison to the sample made from the plain concrete. Almost a similar trend is observed if we compare the compression response of a sample prepared from the cast iron and mild steel. In general, almost all materials, respond identically and will exhibit a stress–strain response on the similar lines; the initial range of the curve can be approximated as a straight line, while its post-peak range can be of any profile—may be a linear or the parabolic, with or without the strain-softening or strain-hardening phenomenon. This all depends upon the microstructure and chemical composition of the material. The confinement provided by the stress conditions also play a major role in deciding the peak load carrying capacity of the material and its post-peak behavior.

2.4.2.1 Parabolic Stress Block

The shape of the stress–strain profile of almost materials under a compression loading follows a parabolic shape. The mild steel, for instance, undergo a linear-elastic response during the initial loading phase, then after reaching its yield point it starts bulging sideway before its failure. From the yield point to its final failure, a set of the large deformations that the sample undergoes are reflected in terms of its post-peak response, generally parabolic in shape caused by the strain-hardening up to certain strain levels and followed by strain-softening. The initial

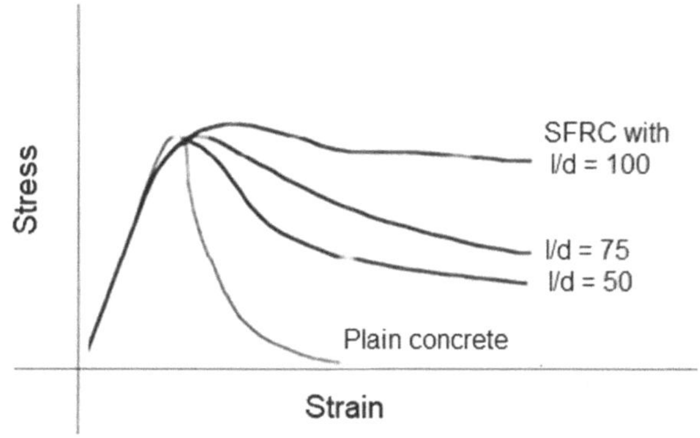

Fig. 2.11 Typical compression stress–strain response of concrete. It also shows the improvements that the steel fibers would make in the concrete response depending upon the fiber aspect ratio (l/d), resulting in the improved post-peak response of the material. *Source* Author

slope of the response curve indicates its modulus of elasticity; it is around 2×10^5 MPa for the steel and the material yield point is not a fixed quantity. It ranges from 210 to 540 MPa depending upon the carbon dosing done at the time of production. On the other hand, the composite materials, like concrete being a heterogeneous material exhibits a response that may end up as a brittle failure or it may be a ductile one; all depends upon the microstructure it possesses in its hardened state. Figure 2.11 depicts a set of typical responses for the plain concrete and those produced using the steel fibers as its internal reinforcement.

The strength of concrete can be determined as the maximum load attained during the loading process under the uniaxial compression conditions. Generally, it is reported as the strength of 150 mm standard cubes at 28 days of the water curing period. The load is applied to the specimen at a uniform strain rate of 0.001 mm/mm per minute. The compressive load normally peaks at an average strain of 0.002 and starts crushing at strain ranging from 0.003 to 0.004. However, it improves to around 0.005–0.009 when the steel fibers are dosed in the concrete mix. The corresponding failure stress, however, improves marginally to $0.75\sigma_{\text{cu}}$ from $0.65\sigma_{\text{cu}}$ (which is normally taken for the plain concrete), where σ_{cu} is the concrete cube strength. Equation 2.2 can be used to calculate the design compressive stress (f_c) of concrete for any value of strain (ε); the partial safety factor of 1.5 is used in the equation.

$$f_c = \begin{cases} 0.45 f_{cu}\left[2\left(\frac{\varepsilon}{0.002}\right) - \left(\frac{\varepsilon}{0.002}\right)^2\right], & \varepsilon < 0.002 \\ 0.45 f_{cu} & 0.002 \leq \varepsilon \leq 0.0035 \end{cases} \tag{2.2}$$

2.4.2.2 Equivalent Rectangular Stress Block

Mathematically, this type of the stress block is convenient to use in place of the parabolic one wherein, the parabolic shape is converted to a statically equivalent rectangular area. The area of the block indicates the magnitude of the compressive resistance load that the member can provide. And, its placement along the beam depth should be such that they must meet the equilibrium condition of $\sum M = 0$ in addition to the other two conditions, i.e., $\sum V = 0$ and $\sum H = 0$, where M, V and H indicates the moment, vertical and horizontal forces, respectively.

2.5 Characteristics Strength of Materials

The previous sections indicate the material strength is highly dependent on its microstructure. All factors that have the capacity to influence or alter the microstructure would definitely tend to modify the material response. The production or manufacturing processes also significantly control the material response and its properties. Any intentional or unintentional change in the material chemical composition and physical dimensions of the final product (e.g., steel rod/reinforcing bar, concrete cube, brick, etc.) will not go unnoticed on the material response that we will observe during the testing. This means that if we test any two bricks or concrete cubes, they are going to fail at the different loads, even if these are made or produced from the same lot of raw materials and even under a similar set of the test conditions (see, the example given at the end of the chapter). An analyst therefore will always get a scatter in the test results. The big question is: how should the test data be converted to a reliable design value? What should be the design load value? Will it be okay to take an average value of the test data? Or, some other statistical approach should be used for this purpose? The answer to these questions is important as the safety of any building or its component is decided by the overall quality of the materials used in its construction and not by one or two lots of materials, which generally have huge variations in the test result data. Therefore a continuous control is necessary to ensure that the strength of any material as furnished is in satisfactory agreement with the value considered by engineers in the structural design of various members and other components.

There are two procedures to approach the best possible solution; namely, (1) Sampling by attributes, and (2) Sampling by variables. For example, a company is manufacturing 1000 numbers of water bottles per day, each having a capacity of 1 L, has decided to randomly select 100 bottles for inspection to measure the compliance of their package weight. Two options are there: one, go for sampling by attributes, and two, go by variables. In the first option, we will randomly pick the bottles and will check whether they are heavier than 1 L or not. That is all, and we will take the decision from the number of bottles that fail to comply with the pre-set acceptance limits for the lot. Nevertheless, this sampling approach is very rough and should not be used for assessing the quality of structural material going to be used in the construction projects. But it can be used for functional materials where the use is not going to influence the load carrying capacity of members. On the other hand, the sampling by variables is based upon their actual weight; it will be noted for each bottle of the sample and the data will be analyzed to see the deviation from the pre-set numbers. This approach provides more information (e.g., if there is a bias or a trend), but is more complex and time consuming. It should be preferred, if the data are taken randomly and independently from the batch and it is expected that the resulting distribution of the data will approximate the normal distribution, also known as the Gaussian distribution; otherwise, the former approach can be used to have a rough idea about the sample quality till the time the actual results start coming from the laboratory.

In 'Sampling by variables' procedure, the analyst needs certain test data from the batch to be examined. The test data points, called as attributes, can be any set of continuously measurable quantities from the batch, such as strength, physical dimensions of specimens, etc. The procedure uses the measured values of these attributes to determine the overall acceptability of the batch. This approach has the advantage of using more information from the tests and therefore is more rational than the previous approach. The material test data and the other physical dimensions related to civil engineering practice generally follow a normal distribution, which is a plot of the probability distribution, always symmetric about the mean value. It shows that the data points near the mean value are more frequent in occurrence than the data points far from the mean value. The probability distribution of any material attributes is illustrated in Fig. 2.12. It indicates that about 68.26% of the data points are falling in an area enclosed between the limits set of the standard deviation, i.e., (data mean $\pm$ 1σ). This area increase to 95.44%, corresponding to the limits, defined by an expression: mean $\pm$ 2σ. This feature of the data distribution is conveniently employed to ensure a desired level of the quality control at the sites.

An acceptable level of the batch quality can therefore be easily defined by means of the data standard deviation (σ) value—Upper limit, defined by an expression: = (mean + σ), and a lower limit, defined by an expression: = (mean − σ). Within these limits, an acceptable quality level can be defined as a maximum allowable fraction of the defective attributes. With an upper limit of (mean + σ), the fraction of the non-desirable values in the data set is equal to the area under the distribution function to the right of this limit. This value is then compared to the allowable fraction to determine the acceptability of the batch. Similarly, the values on the left of the lower limit can be determined. Alternatively, the limits (both, upper and lower) can be decided by starting from any desirable value of the sample mean value.

Using this procedure, we can easily set a target of achieving that value of the desirable probability of occurrence of the value from the sample mean. Characteristic value of the attributes is usually defined to achieve this objective. For example, the characteristic strength of the material is usually adopted in the design of any structural element instead of an average or some other value of the data. It is defined as the strength of material below which not more than 5% of the test results are expected to fall. This means that the mean value of the material strength has to be significantly greater than the 5 percentile characteristic strength considered by engineers in the design process of structural members. Physically, it indicates the resistance that any material will provide to the load below which the chances of its failure are only 5%. For example, Concrete having characteristic strength of 25 MPa means if a 25 MPa stress or low is developed in it, the probability of its

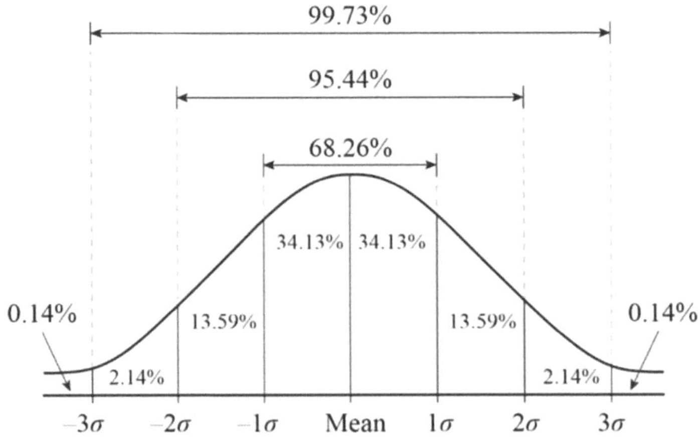

Fig. 2.12 Typical set of data that follows a normal distribution. *Source* Author

failure is only 5%. This strength parameter forms the backbone of the reliability based structural design. A typical plot of the probability function of some data related to the material strength is shown in Fig. 2.13. We can easily achieve this target if we adopt the upper and lower value of the acceptable limits for the test data as (mean $\pm$ 1.64σ).

For example, if we want that the bricks to be used in any construction work must have a crushing strength of 10 MPa or more; then, from the known value of the sample standard deviation, σ (say, 1.5 MPa), the characteristic strength of bricks can be easily determined as 12.46 MPa (= 10 + 1.64 $\times$ 1.5). Alternatively, the mean value of the material strength (bricks in the present case) can be calculated from the desired characteristic value to be considered in the design process. In such a case, the average strength of the bricks from the test data must be in conformity with this calculated value to ensure that about 95% of the test data must have strength values equal or more than 10 MPa.

Mathematically, let n be the number of specimens in the sample and x_i, $i = 1, 2, 3, \ldots, n$, be the measured values of the test attributes. The standard deviation (σ) of the sample having a mean of x_m can be determined from Eq. 2.3.

$$\sigma = \sqrt{\frac{\sum (x - x_m)^2}{n}} \tag{2.3}$$

It is important to note that Eq. 2.3 should be used only when the standard deviation (σ) is computed using all values in the population. For instance, if we have to ensure quality of the total concrete used in some project, it is not possible to calculate standard deviation for the entire work unless its test results are available from the total concrete used (or going to be used) in the project. Some test results may come after one month and some may be after the six months when the

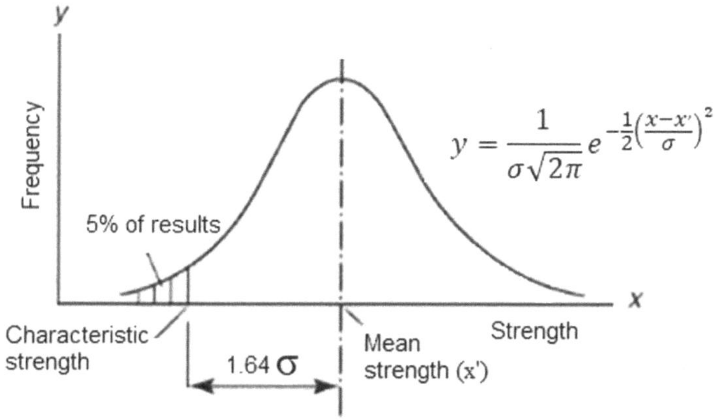

Fig. 2.13 Characteristic strength of materials from the normal distribution function of some typical test data. *Source* Author

respective concreting job is complete. In such a case, site engineers normally take decisions by computing the standard deviation from a small sample out of the entire population (of test data that will be available after a long time when the corresponding work event is completed). It is always recommended in such cases to take the denominator (n) in the Eq. 2.3 as ($n - 1$), which will provide an unbiased estimate of the true population variance.

If the standard deviation (σ) and the mean value (x_m) of the test data (the attribute can be the strength, load, or specimen dimensions or some other physical parameter) is known, Eq. 2.4 can be used to determine its characteristic value (x_{ck}). If we are checking the load carrying capacity of a set of standard concrete cubes (say at 28 days), the variable (x_{ck}) will be the compressive strength of cubes, called as the concrete characteristic strength (f_{ck}) at 28 days. It will ensure that about 95% of all test cubes posses compressive strength of value more than the value (x_{ck}).

$$x_{ck} = x_m - k\sigma \tag{2.4}$$

The value of the factor (k) in Eq. 2.4 depends upon the sample size (n). The value of the factor can be taken as 1.64 for an infinitely large sample of random independent variables of size, n (it may represent the entire population), and the other samples, its value can be taken from the Table 2.1 depending upon the sample size (n).

Table 2.1 Value of the factor (k) for different number of specimens

Specimens, n	1	2	3	4	5	6	8	10	20	30	∞
k	2.31	2.01	1.89	1.83	1.80	1.77	1.74	1.72	1.68	1.67	1.64

Data source [12]

Most of the design and quality control guidelines specify the value of the factor (k) as 1.65, approximating a sample size of around 30 specimens. Normally, the number of specimens that should be taken from the batch or shifts is fixed by the guidelines and other applicable code of practices depending upon the quantity of the material used at the site. For instance, it is specified as one sample of three specimens for 1–5 m^3 of concrete used at the site; it is two samples for 6–15 m^3 and four for concreting of 31–50 m^3. It is therefore always better to consider the denominator (n) in Eq. 2.3 as ($n - 1$) to remove the bias in the variation that is likely to be there in case of the small sample size. Similarly, different guidelines specify the number of samples that must be taken randomly from the entire batch of materials, may be bricks, steel rebar, soil, wood, metals, etc. to determine their acceptance. Alternatively, form the available quality of the materials to be used in some project; we can determine their characteristic strength to be used later in the design of products, structural members, etc. so that they could safely perform their basic task of the load transfer in the given structural system, albeit it may come at the higher costs as the material consumption will increase if the strength is low because of poor quality materials. It is only the skill of the site engineers that how they are approaching the problem and handling the situation to optimize the project costs.

Illustrations

A set of thirty standard concrete cubes have been sampled randomly from a construction site, and these are tested at 28 days. The compressive strength of these cubes is tabulated below:

Cube No.	Compressive strength, MPa	Cube No.	Compressive strength, MPa	Cube No.	Compressive strength, MPa
1	24.89	11	20.00	21	21.78
2	30.22	12	21.33	22	24.44
3	25.78	13	24.89	23	25.33
4	26.22	14	25.33	24	26.67
5	23.11	15	26.67	25	28.89
6	22.22	16	28.89	26	31.11
7	21.33	17	30.22	27	17.33
8	27.11	18	31.11	28	27.56
9	23.56	19	22.67	29	29.78
10	24.00	20	23.11	30	19.11

What should be the safe value of the compressive strength that can be taken in the design of the concrete member for this concrete lot?

The safe value is the strength that the entire lot of the concrete will be able to provide. It will not be an average value, which indicates that 50% of the tested values are coming more than that value while the rest will come less than it. This property of the average value can be easily verified from the tabulated results. The average value (= the ratio of the sum total of all values and the total number of cubes in the sample) for this case is 25.15 MPa, and around 15 cubes have a strength of more than this average value. Any member designed using the average value of the strength will be unsafe to support the given load as the strength of the material will vary along different pour locations in the member.

The characteristics strength (f_{ck}) will be a better choice to compute the safe value of the cube strength. Equation 2.4 can be used for this purpose that would require mean value (x_m) of the sample strength as well its standard deviation (σ).

The mean strength of the sample containing thirty cubes (n = 30) = 25.15 MPa

Using Eq. 2.3 with the denominator of ($n - 1$) gives standard deviation = 3.62 MPa

The value of the factor (k) can be taken as 1.67 for a sample of thirty cubes (see, Table 2.1).

$$f_{ck} = x_m - k\sigma = 25.15 - 1.67 \times 3.62 = 19.11 \text{ MPa}$$

This indicates that we can adopt a safe value of the concrete strength as 19 MPa, which will ensure that only 2 cubes (= 5%) in the sample have a load carrying capacity of less than 19 MPa. We have only two such values, namely 17.33 and 19.11 MPa in the sample and all the rest possess strength to support a load corresponding to the characteristic strength.

Bibliography

1. Brandon D, Kaplan WD (2008) Microstructural characterization of materials. John Wiley & Sons Ltd, West Sussex
2. Brundle CR, Evans CA, Wilson S (1992) Encyclopedia of materials characterization—surfaces, interfaces, thin films. Butterworth-Heinemann, Stoneham, MA
3. Choo B, MacGinley T (2018) Reinforced concrete: design theory and examples. CRC Press, UK
4. Courtney TH (2013) Mechanical behavior of materials. McGraw-Hill, London
5. Gleiter H (2000) Nanostructured materials: basic concepts and microstructure. Acta Mater 48 (1):1–29
6. Hull D, Bacon DJ (2011) Introduction to dislocations. Butterworth-Heinemann, Oxford
7. Mehdi Ashraf S (2017) Practical design of reinforced concrete buildings. CRC Press, United States
8. Pillai SU, Menon D (2009) Reinforced concrete design. McGraw-Hill Education (India) Pvt Limited, India

9. Shackelford JF (2009) Introduction to materials science for engineers. Pearson Education, New Jersey
10. Tetelman A, Barrett CR, Nix WD (2005) The principles of engineering materials. Prentice-Hall, Englewood Cliffs, NJ
11. Weidmann G, Lewis P, Reid N (1990) Structural materials. Butterworth, London
12. Singh H (2017). Steel fiber reinforced concrete: behavior, modelling and design. Springer Transactions in Civil and Environmental Engineering, Singapore

Chapter 3
Characterizing Material Response

Materials form the backbone of any engineering project. Whenever any idea or conceptual design is converted to some working model and/or prototype, the role of materials is of utmost importance; rather without them, it is not possible to construct any facility, machinery, vehicles, buildings, bridges, roads, canals, etc. Engineers identify the need to have something, maybe bridge, vehicle or some building, and then work together as a team to develop some conceptual framework based on the fundamental engineering principles, on which all such physical entities are built by using the materials appropriately as per the design and identified specifications. There is a long list of materials, both natural and man-made, at their disposal. They have to select the most appropriate one depending upon the need of the project and cost considerations. This selection is mainly made on the basis of various strength properties and other factors, such as their load carrying capacity, the toughness, the ductility and durability; unit weight; constructability; and the fire resistance. The process by which the structural and other material properties are probed and measured is termed as material characterization. It is a fundamental process in the field of materials science, without which it is not possible to develop any scientific understanding of the response and other properties of engineering materials. This becomes the most important part of any project planning and design process as all materials depending upon their type and the microstructure respond differently to various physical disturbances, such as mechanical loading, thermal loading or any other cause that tends to either elongate or shorten the substance or any object made up of all such materials; thereby, it is going to be the major influential factor among other factors that would control the structural design process. A precise and accurate determination of the material properties, therefore, is the first step in the selection process and their efficient utilization through design in various components of any project. Different ways and means to characterize the material response are described in the following sections:

H. Singh, *Structural Materials*, Materials Horizons: From Nature to Nanomaterials,
https://doi.org/10.1007/978-981-16-3211-2_3

3.1 Different Equipments and Devices

It is all natural for the materials to deform when they are physically disturbed. For example, a solid object undergoes change in its length under the action of an axial force; it may shorten or elongate depending upon the nature of the applied force. Some of them may deform to a significant extent, while some others may do so negligibly. Materials depending upon their chemical composition and the microstructure mobilize their internal resistance to oppose all such deformations. The maximum value of the resistance offered by them just on the verge of their failure is called as strength. Therefore, in order to mobilize this full resistance, they may have to deform considerably; materials go on deforming as the load is increased and they exhibit a load–deformation response during the process, normally very unique to each and every type of the material. So, an analyst is mainly interested to have an experimental setup that should be able to capture the deformations being exhibited by a material or some object made up of all such materials during the loading process and also, the magnitude of the force imposed over the said object or material. Such a test setup is usually equipped with some set of devices to perform the identified tasks. A device is any specialized part or component of equipments attached to perform a particular task. These are usually mechanically or electrically operated accessories but form an important part of the main equipments.

3.1.1 Force Measurement

Force is a measure of the interaction between bodies. It can be between two or more bodies as small as atomic particles and as large as galaxies in the universe. But from the engineering point of view, it is between two or more structural elements or objects, such as the slab and beams; the column and footings; beam and another beam; mass of some user and structural member, etc. It is a physical quantity that flows seamlessly through the bodies unless these fail to provide a safe passage to it (and that depends entirely upon their capacity to bear the load). During this process, it tends to give either pull or push effect on the object if the force is applied along the member-axis. On the other hand, if the force is acting normal to the member-axis, it will start giving bending effect on the member.

It is worth to recall, that the mass is a source of all forces that operate between two or more bodies. It is directly linked to the number of atoms that make up the bodies; therefore, it does not change with position and movement unless some material is added or removed from the object. On the other hand, the force will change with the change in the distance between the bodies. Newton's law of universal gravitation governs the magnitude of all such forces that would induce between the two masses placed at some distance apart. The unit of mass in the SI system is the *kilogram* (kg) which is equal to the mass of the international prototype 'kilogram' held at the International Bureau of Weights and Measures in Paris.

Whereas, the force is expressed in Newton (N) which is a measure of the force needed to give an acceleration of 1 m/s^2 to a mass of one kilogram.

Being a vector quantity, the force has both direction and magnitude. So, if the forces acting on a body in equilibrium are summed, then they always add to zero. This means that if their sum total is not coming zero, then the body is not in equilibrium, maybe 'temporarily,' and the body will start accelerating such that the rate of change of its momentum will generate enough force to maintain the equilibrium conditions, albeit in the motion. However, if it is held in position by some means, the supporting structure in all such cases will provide the desired reactive force to maintain the equilibrium of the system. Irrespective of the condition considered, the body always deforms. This condition forms the basis of most of the force measurement systems currently in use and provides us an opportunity to measure the magnitude of the force acting or applied over some object.

The simplest way to measure the force required to crush any material therefore is to just place some mass over the top face of a test specimen in increments till it fails to hold any additional load. The resistance offered by the material to the deformations maintains the equilibrium conditions within the system; the highest resistance offered by the specimen at the failure is a measure of its strength.

Alternatively, the structural and the piezoelectric properties of certain materials can be used to measure the strength of test specimens. In the former case, the load–deformation responses of certain system are used to measure the amount of force imposed over the test specimens during their testing. For instance, a spring balance or proving ring can be easily used for this purpose if their stiffness (= ratio of the force to the deformations produced by any such force, called as the device constant) are known. Normally, this type of the information is supplied by the device manufacturers in the form of a linear relationship or in the tabulated form. So, by noting the deformation in the specimen corresponding to certain load increment, the value of that load (increments) imposed over the specimen can be determined by multiplying the deformation so produced in the specimen and the respective device constant. These two methods are purely mechanical in nature as an analyst has to physically note down the deformations being exhibited by the specimen at each load increment, either by means of a scale/ruler or the dial gauges used to record the deformations and finally, convert this data to the strength related properties. Figure 3.1 illustrates the commonly employed proving rings and spring balance that can be used in different test setups.

Certain materials posses piezoelectric characteristics by which they produce electric charge on its opposite faces in the case of any mechanical deformations imposed on them. There exist a unique relationship between the input (stress or strain imposed on the material) and the output (electric charge producing a potential-difference across its thickness) for each and every type of piezoelectric materials, known as response curve. These types of materials find enormous applications in the manufacturing of various sensors. So, if the response curve of certain sensor, say load cell, is known (usually, it is supplied by the manufactures), the magnitude of the load applied to the material test specimen during its testing can be easily determined. Figure 3.2 depicts a typical load cell along with its vital

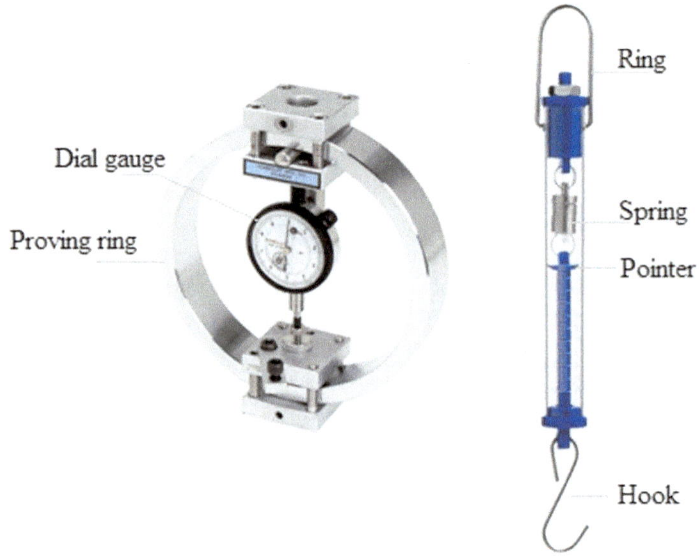

(a) (b)

Fig. 3.1 Typical depiction of **a** proving ring fitted with a dial gauge to measure the deformation, which subsequently will be converted to the load producing such deformations, **b** spring balance in its uploaded position. *Source* Author, Heavy Testing lab, GNDEC

(a) (b)

Fig. 3.2 Typical depiction of **a** load cell, **b** Strain gauge that consists of long thin wire (black line) arranged in a grid shape. *Source* Author; Heavy Testing lab, GNDEC

components. During the loading process, the test specimen undergoes deformations, and if certain load cell is placed above the specimen, it will also experience the same set of deformations; thereby creating electric signatures corresponding to each load increment. From the response curve of the device and/or load cell, these signatures can be converted to the load being imposed over the specimen.

Another type of the load cells works on electric properties of materials, the major among them is their resistance characteristics. These types of cells are called as strain gauge load cells. Due to their accuracy, versatility and cost-effectiveness, these are more common than any other type of the load cells generally used in various industrial applications. Instead of the piezoelectric characteristics of the materials, they rely on the electric properties of the spring elements used in the body of the load cells. The strain gauges attached to the body of the load cells consist of very thin wire or foil, arranged in a grid pattern (see, Fig. 3.2c) over the flexible backing, which enable it to be easily applied to a load cell, mirroring the minute changes to be measured during the testing. Mechanical deformations imposed on the load cell change the shape of the strain gauge, which in turn results in change in electrical resistance of the gauge. A tension force stretches a strain gauge, causing it to get thinner and longer, resulting in an increase in the resistance; whereas the compressive force produces an opposite effect. It is important to mention that usually a set of resistances has to be used in the cell to measure accurately the small change in the voltage across the circuit caused by the change in the resistance of the element during the deformations.

The Wheatstone bridge, for instance, consists of a set of four resistors to pick up the signal; these are arranged in a specific order as shown in Fig. 3.3 that enable the analyst to measure the even small changes in the voltage across its terminals, whenever any of these four resistors undergo alternations. This type of electric bridge is pasted over an internal elastic body placed inside the cell. Whenever the cell is loaded by external forces, this part deforms and also causing the strain gauge fixed therein to deform, which leads to some change in their resistance responsible for the voltage fluctuations observed across its terminal.

Like the piezoelectric materials based load cells, there exists a unique relationship, known as response curve, between the cell output, usually measured in terms of the voltage change, and the corresponding deformations imposed to the sensor fitted in the cell body by the external applied load. This curve can be used to measure the load applied to the test specimens. It is, however, utmost important to avoid any overloading of the load cells as it may permanently distort and damage

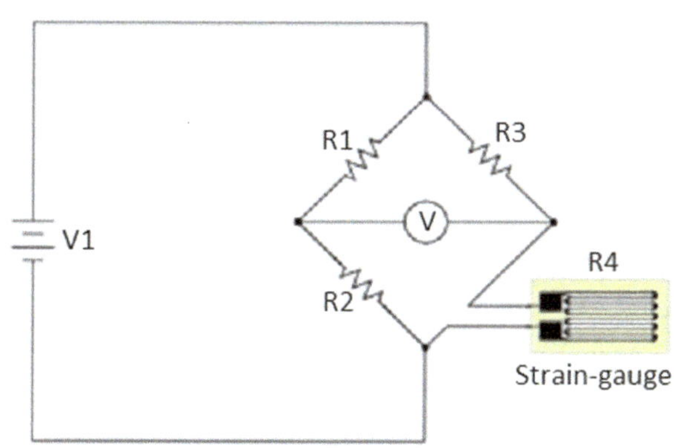

Fig. 3.3 Typical circuit layout of the Wheatstone bridge used to convert the mechanical deformation to an electric signal, usually the voltage change, V_1 in the form of resistance R_1, R_2, etc. *Source* Author

the strain gauges fixed over the elastic body of the cells. The routine checking and calibration of the load cells is thus a very important aspect of the equipment/device maintenance schedule. Otherwise, the probability of getting erroneous results from the use of such devices during the testing will be high and the test results in such a case may show a larger than accepted variations/scatter in the data.

3.1.2 Strain Measurements

Strain and the stress in any deformable body always coexist in the state of the equilibrium. Actually, both are the outcome of a phenomenon called as 'causality dilemma,' wherein each parameter is the result of the other. Whenever any deformable body is loaded, it undergoes shortening or elongation depending upon the type of the stress imposed. The extent to which it deforms is the strain, which mainly depends on the elastic modulus of the material of the body, whereas the stress is a measure of the material's resistance to the deformations. A body having a higher modulus exhibits small deformations, and vice versa, and accordingly, it may need a large and small force magnitude, respectively to cause any such deformation in the body. It is important to note that the source of straining can be a thermal loading or it can be some mechanical loading.

The change in the dimensions that the body undergoes during the loading process can be easily measured by means of a ruler. Most of the commonly available rulers have a least count of around 0.5 mm. So, these can be used only if the body is loaded to an extent that the resultant deformations are at least of this significance value or in all cases, where any small error in the measurement is not going to affect the end result. For instance, consider the case of the materials (like, rubber) having a low modulus; these will undergoes large deformations even under the action of a small magnitude of the applied load, or another case, where the load magnitude is large enough to produce deformations that are measurable with the desired accuracy even in otherwise stiff materials, the ruler is the best and cheat option of measurements in all such scenarios. If the use of rulers is not helping to obtain the desired accuracy in measurements, then, there are many other devices that allow taking measurements up to an accuracy of 0.001 mm. Vernier caliper scale, Vernier micrometer and screw gauges are the few examples from this category of devices. These are available both in the digital as well as analog variants. Figure 3.4 shows a typical set of commercially available models of the Vernier calipers. Similarly, micrometers are another set of devices capable of measuring linear dimensions up to a range of microns (0.0001 mm).

In addition to the lower least count, another major drawback of the ruler is that it can be used to take only one measurement at a time, e.g., if we are testing a mild steel specimen under an axially applied compression load, it will shorten its length as well as undergoes an increase in the diameter due to the lateral bulging effect. So, if we are interested to determine the change in the length or the diameter of the test specimen, the only option is to pause the loading and take the measurement. A good

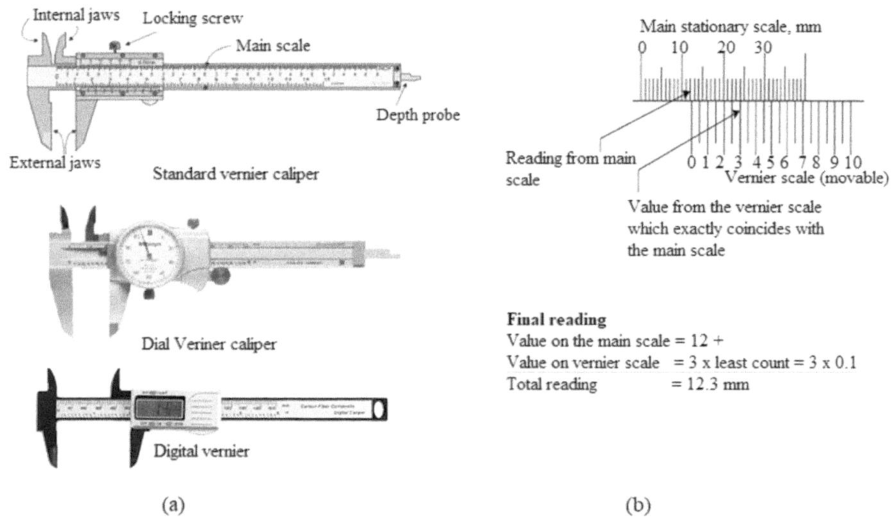

Fig. 3.4 Vernier Caliper: **a** different types, **b** illustration to take a measurement on the Vernier. *Source* Author

option, however in this case would be to use a Vernier caliper as its internal and external jaws would allow the user to take the measurements of changes that occur in the height and diameter of the test specimen, respectively, at any interval of time. Once the change in the dimensions of the test specimen is recorded, it can be easily converted to the corresponding strain values by fractioning it over its original dimensions. Similarly, the maximum change in the length of a test specimen that takes place under a tensile loading, known as its elongation and usually expressed as the percentage change in the length, can also be determined using such devices.

However, if the purpose of investigations is to continuously monitor the deformations being exhibited by a test specimen under some incremental loading, we need something that should enable continuous measurements of the changes that are taking place in the test specimen during the loading process. Extensometers and the linear variable displacement transformer are the type of the measurement devices that permits such type of continuous monitoring. They are commonly known by their abbreviation 'LVDT.' Figure 3.5 depicts a set of LVDTs, which works by converting the linear displacements being sensed by them into an electrical signal through the principle of mutual induction. These can be used to measure the linear movements in the range of 0.25–1000 mm.

The moving arm of the device can travel only along its axis and the extent to which it moves from a reference point is captured by the LVDT as its output signal (V). Basically, it works on the principle of an electrical transformer. Figure 3.5b shows a typical arrangement of the internal coils used in a typical LVDT sensor. It consists of a coil assembly and the core. Being attached to the body of the LVDT, the coil assembly therein remains stationary; while its core can travel freely inside

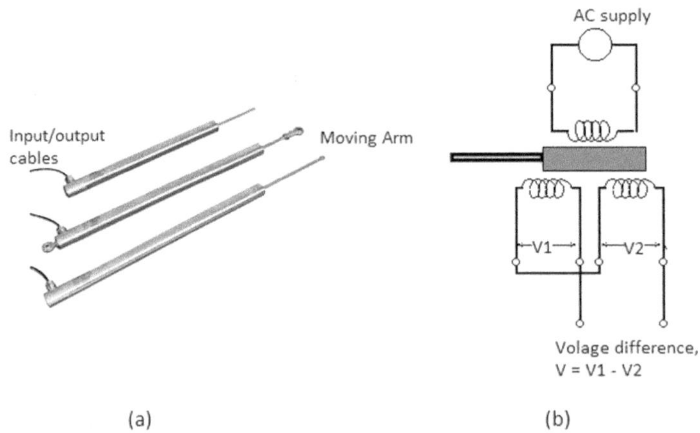

Fig. 3.5 Linear variable displacement transformer: **a** different types, **b** its working principle depicted through a typical internal electrical circuit. *Source* Author

the coil assembly of the LVDT sensor during the course of taking the measurements. The core of the sensor is made using the magnetically permeable materials; while its moving arm is made from the non-ferromagnetic materials to minimize/avoid the interference from the sensor body on the measurements.

The coil assembly consists of a set of three coils—a primary and two secondaries, generally wound on a glass-reinforced polymer hollow shaft. These coils are arranged over the shaft such that the primary coil is located between the two secondary coils, generally wound in the series, but in opposite directions as is shown in Fig. 3.5b. This type of the coil placement is done purposefully to get the desired outcome from the sensor. When the AC voltage is applied to the primary winding, it induces voltages in each of the secondary windings. However, the resulting flux in these coils is coupled through the intermediately placed core. The non-ferromagnetic nature of the arm of the core changes the flux that induces voltage in the secondaries as it moves in the assembly. The resultant differential voltage (V) in the secondary coils determines the distance moved by the arm during the measurements, and the phase of the voltage indicates the direction of core movement. Since there is no physical contact between the fixed and moving parts of the device, these are extremely robust and versatile in nature and are capable of working in a wide temperature ranging from cryogenic temperatures to as high as 650 °C, in harsh environments, and even under the high vibrations and shock levels.

Extensometers are another set of low cost and easy-to-use devices that permit measurement of the changes that takes place in the length of test specimens during the testing operations. Two variants of the extensometer are available that can be used for different applications, namely the contact type and the non-contact type. It is possible to fit the contact type on the specimen itself, in some specific location, to monitor the change in the length taking place in that region of the test specimen;

these are also called as the clip-on extensometer. The mechanical version of this type of extensometer allows measures in a range of 1–100 mm; whereas, its digital version comes fitted with LVDTs as their 'sensor arm' to take the readings. Both versions of the extensometer whether mechanical or digital enable capturing the readings continuously as the specimen is loaded. Unlike the mechanical version of the extensometer, the digital version of the device allows the measurements until failure as they are fitted on the supporting system of the test machine and not over the specimen as in the first case. They are capable to measure elongations as small as few micrometers and up to 1000 mm without losing any accuracy. On the other hand, the non-contact type extensometer uses a laser beam to observe the changes that take place in the test specimens and uses certain image processing techniques to convert these changes to strain or displacements. Unlike the other types, they are also capable of taking measurements as small as one micrometer. Figure 3.6 shows the different types of extensometers that are commonly used in the routine testing.

While the extensometers and the LVDTs measure the displacements as the test specimens deform during their testing operations, there is another range of the devices that are capable of measuring directly the strains the specimens will experience during the loading process. These uses piezoelectric patches or a set of resistors in the form of the Wheatstone bridge to capture the changes in dimensions of a test specimen (See, F.ig 3.3); these are known as strain gauges. The working of these devices is almost the same as is used in the load cells to measure the force applied over a test specimen or the resistance offered by the materials to the imposed deformations; but instead of reporting the load values as their output, the strain gauge patches are calibrated to display their output in the form of changes that occur in the dimensions of the test specimen during the testing operations. A typical patch of the strain gauge along with its fixing arrangements on a test specimen is depicted in Fig. 3.7. The patch is pasted on the surface of the test specimen using some permitted adhesive. The manufacturers of patches usually prescribe a range of adhesives that can be used to achieve the desired accuracy in the results. The adhesive generally used for this purpose are Duco cement (cellulose

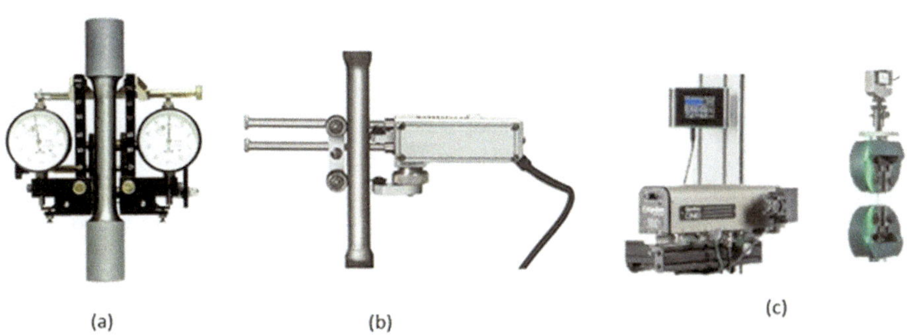

Fig. 3.6 Extensometers: **a** contact type—mechanical, **b** contact type—digital, **c** non-contact type. *Source* Author

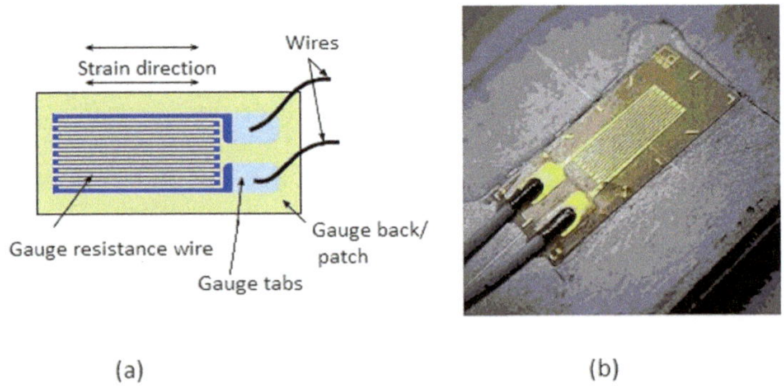

Fig. 3.7 Strain gauge: **a** parts, **b** fixing of the strain gauge over some test specimens. *Source* Author; Heavy Testing lab, GNDEC

nitrate), bakelite cement, epxoy, cynoacrylate cement and ceramic cement. Among these adhesives, the cynoacrylate cement is unique as it does not need heat or other catalyst to start the polymerization leading to the bonding process. On the other hand, for high temperature applications, ceramic cement can be one of the choices.

The user has to be very careful while fixing the gauges over the test specimen. The strain gauge patch should be fixed along the direction of a possible strain measurement as it can monitor the change in the length or in the width of a test specimen only along its marked axis; a wrong pasting on the test specimen can result in a permanent damage to the strain gauge patch and non-acceptable error in the observed value of the measurements.

Dial gauges are another important device generally used to measure the linear movements being exhibited by the loaded specimens. These are commonly used to measure the deflection of a loaded test beam specimen and also, in the proving rings, as well as many other situations where a small measurement need to be registered or indicated. Dial indicators typically have a least count varying from 0.001 to 0.01 mm and these are available to measure displacements ranging from 0.25 to 300 mm. These are very easy to use and like the LVDTs, these can be fitted on the stands while taking the measurements. Figure 3.8 shows a set of commonly used dial gauges.

Dial gauges are the simplest device to measure the linear displacements. The analyst can record the single reading at a time or they can take continuous measurements of the deflections, settlements, indentation, etc. caused by the incremental load or otherwise also. The knob of the gauges is touched at the monitoring point in the member under the investigations. The knob can move up and down as the member deforms and the extent of the movement is reflected in the gauge reading, which can be converted to the deflection or settlement value by multiplying it with the device constant (= its least count), e.g., if the reading being shown on the gauge is 125 (an absolute number), it will translate to a corresponding displacement of 1.25 mm (assuming a device constant of 0.01 mm), and it will be 0.125 mm if we have used gauge with the device constant of 0.001 mm. An appropriate dial

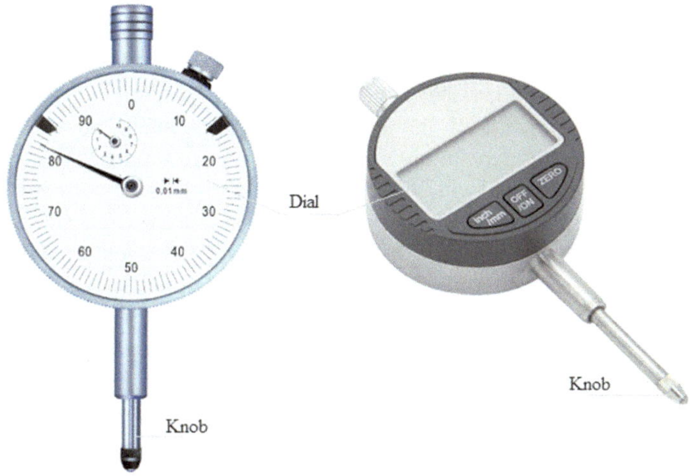

Fig. 3.8 Dial gauge: these are available both as mechanical (see, **a**) and digital variants (see, **b**). They have a knob that can move in-out as the body deforms during its loading; this movement of the knob is displayed on the dial as an absolute value. By multiplying this value with the device least count, it can be converted to the displacement value. *Source* Author

gauge should be selected depending upon the value of anticipated deformation on the object. For instance, if we need record very small deformations or otherwise also when we need to observe up to say three decimal levels, the dial gauge with a least count of 0.001 mm will serve our purpose in all such cases.

3.1.3 Displacement-Controlled Testing

While determining the response of any loaded material, maybe tension or compression type of the loading conditions, we need some arrangement/setup where the test specimen could be placed. There must be some system in the setup to apply the load over the test specimen and record the corresponding measurements related to the deformations being exhibited by the test specimen during the loading process. In a displacement-controlled test setup, the test specimen is continuously strained at a pre-set rate till it fails. The loading strain-rate depends upon the type of the material being tested and the purpose of the testing operations; it can be kept slow to simulate a gradually applied loading conditions or it can be fast for an impact type loading condition. It is worth to recall, that the loading rate greatly influence the material response and its strength properties. It is therefore utmost important to be very careful to select the appropriate strain rate while capturing the response of materials.

Whenever any material is strained, it tends to resist the imposed deformations by mobilizing the internal resistance; it is all natural. If the strain is small, it will offer accordingly smaller resistance and it will go on increasing as the test specimen is progressively strained under the action of imposed incremental loading. The response exhibited by most of the materials remains linear up to the limit of proportionality and thereafter, it may vary from material to material—it is generally a nonlinear in nature with or without strain-hardening or strain-softening behavior. Depending upon the type of the materials, some of them undergo strain-hardening while others exhibit strain-softening response. The displacement-controlled testing allows capturing precisely the complete load–deformation response curve of such type of test specimens.

The specimens can be loaded either using an electromechanical loading arrangement or by means of a hydraulic system. In the former system, a variable speed electric motor is used along with a gear reduction system to move the cross-head up or down on the supporting ball-screw drive system, generally consisting of one, two or four screw-towers. In the routine machines of rated capacity up to 2000 kN, the supporting arrangement consisting of two screw-towers are often used. The movement of the cross-head loads in the screw-tower can manifest as a compression force or tension force depending upon the placement of the test specimen between the jaws. By changing the speed of the motor, a variety of cross-head speeds can be achieved. Whereas, a hydraulic system based machine uses either a dual or single acting piston or ram to move the cross-head up or down. Unlike the electro-mechanically operated testing machines, the operator manually adjusts the orifice of a pressure compensated needle valve to achieve the required strain rate. However, a precise movement and control of the cross-head to any desired accuracy can be achieved by employing the microprocessor based closed-loop servo system. Nowadays, almost all testing machines come fitted with closed-loop servo system for controlling the testing operations. Different loading arrangements can be used to load the test specimens in tension, compression, bending as well as in shear just by fitting and using an appropriate set of grips in the machine to hold the test specimens.

The testing machine uses a set of the load cells, LVDTs and other sensors to take the various measurements related to the imposed strains and the resistance being mobilized in the specimen. A schematic representation of various components of testing machine is shown in Fig. 3.9. The test specimen is gradually strained at the pre-set rates. The test specimen usually generates a unique pattern of the internal resistance to oppose the imposed deformations; it all depends on the type of the material used to prepare the test specimens. It is worth to note that the strains being imposed on the test specimen are continually increasing during the loading process, right up to the point when the specimen fails to hold any further load. The load cell fitted in the machine continuously records the resistance being offered by the test specimen; it may increase or may reduce or may remain unchanged depending on the nature of material. At any imposed strain level, if there is an increase or reduction in the resistance, may be because of initiation of some internal cracking or any change in the microstructure of the material, it will be recorded by the load cell

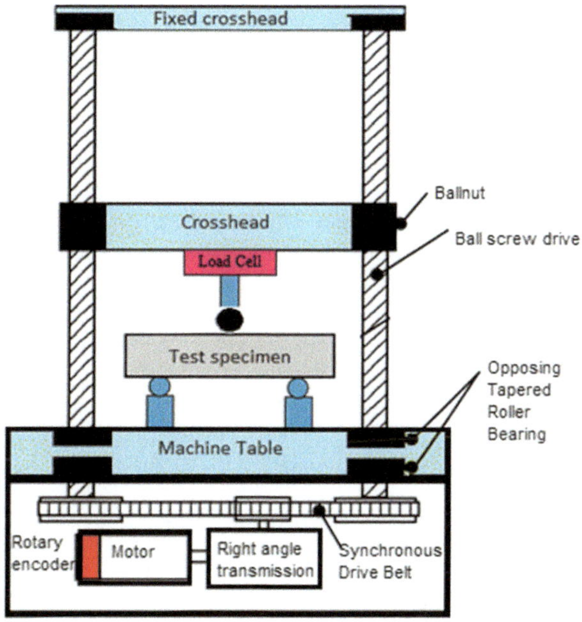

Fig. 3.9 Schematic presentation of different components of loading arrangement and data capturing system of displacement-controlled testing machines. *Source* Author

as such and it will be reflected in the response. As a result, the analyst is able to capture a complete load–deformation profile of the material right from the start to the point when the test specimen fails to hold any additional load no more.

3.1.4 Load-Controlled Testing

In the load-controlled testing machines, the movement of the cross-head is directly linked to the rate of load application, applied either hydraulically or by means of a mechanical system driven by electric motors. This means that the cross-head in this case moves at a fast pace in comparison to the displacement-controlled machines, with each step equal to the magnitude of the applied load increment. The corresponding deformations being exhibited by the test specimen are recorded using either dial gauges, or LVDTs or any other similar device. As this type of the loading arrangement cannot capture the material resistance being mobilized by the test specimen in response to the external deformations, the analyst will not be able to precisely plot the load–deformation response of the loaded specimens, especially their post-peak response, where the specimen resistance fluctuates up and down a number of times. The overall response for most of the materials in this range becomes nonlinear in nature.

Unlike the displacement-controlled machines, the rate of applying the load increments is controlled in these machines such that their magnitude per unit of time remains constant throughout the loading process. As pointed out earlier, during the loading process, especially in the post-yield/cracking point, the resistance of a specimen increases or reduces a number of times depending upon the material used to prepare test specimens; the chances of missing such an important data point therefore, are very high in the case of the load-controlled machines.

In these types of machines, as the purpose of the loading arrangement is to just apply the load at some predetermined rate, so naturally, the load cells are not used here to monitor the load application. Instead a set of load gauges is fitted in the machines for monitoring the amount of the load applied to the specimen. It also reduces the cost of the machines. As such, the applications of these machines are limited only to the determination of the strength of the materials and/or the percent elongation. These cannot be used to determine the resilience or toughness of materials, which needs a complete profile of the load–deformation response of the test specimens.

3.1.5 Data Acquisition System

Irrespective of the machine used in the testing of specimens, a lot of data is generated during the entire loading process. This data mainly contain points related to the deformation imposed on the test specimens and the corresponding value of the resistance mobilized by the specimens; or alternatively, the data set may contain load increments and the corresponding value of the deformations being exhibited by the test specimen. This data set is a bare minimum output from the machines. Additional information in the form of strains, crack widths, etc. may also be there if some extra sensors are attached to the test specimen to capture these data points. Generally, the data will be available in the form of electrical signals as an output from the sensors fitted in the machines or those obtained from the sensors attached separately to the test specimens.

Data acquisition is the process of sampling these signals that measure the real world physical conditions of the test specimens and converting the resulting samples into digital numeric values. The information so collected is generally processed using the computer to draw some useful inferences from the test data. A complete data acquisition system consists of a set of sensors (load cells, LVDTs, strain gauges, etc.) and actuators generally fitted into the testing machine along with a signal conditioning circuitry, analog-to-digital converters and a computer to process the data. A typical layout of these components is shown in Fig. 3.10. Each component has a definite role to play in the system: Sensors convert the deformations or the loading imposed on a test specimen into the electrical signals. These signals may come with a lot of the noise and unnecessary data that can be filtered out without affecting the accuracy and the purpose of the testing; it also helps to reduce the size of the data; this process is performed by a signal conditioner. The analog-to-digital converters in the system transform the conditioned sensor signals to digital values, which the computer will read and use to process the data as per the requirement of the users.

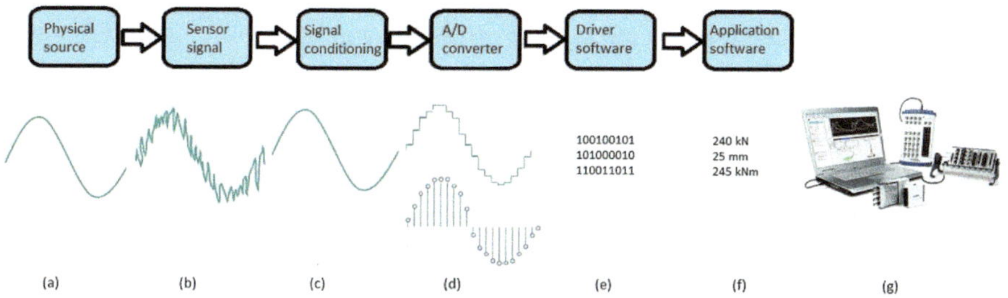

Fig. 3.10 Data acquisition system (DAS): **a–f** block diagram illustrating different functions that DAS usually performs. It also indicates a typical physical disturbance (**a**) and how sensors normally pick up that signal, with a lot of noise (**b**), followed by conditioning of the signal to an almost smooth curve by filtering out the noise (**c**), and subsequently, its conversion to a digital signal that the computer will process for data interpretation; (**g**) a typical depiction of the data acquisition system. *Source* Author

The data acquisition system is a mandatory component of the displacement-controlled testing machines. These machines are usually programmed through the in-built data acquisition system to display their output in a graphical format depicting the load–displacement response of the test specimens being studied. Depending upon the other sensors attached or used during the testing, the machines can also display the variation of the respective output signal, such as crack widths, temperature, and strains, etc. against the strain or the load values plotted as the abscissa of the response curve.

3.2 Selection of Suitable Testing Equipments

Structural materials deform under the action of any mechanical disturbance may be thermal or otherwise so do the members made from such materials. The prime purpose of the testing materials is to get their strength, toughness, stiffness and load–displacement response. Their durability characteristics are also required sometimes to see how they would perform over a period of time. The knowledge of these parameters helps the engineers select appropriate materials for the given design constraints and then, proportion the structural members using their safe stress values to perform the intended functions.

Each equipment and sensor come with a set of their performance limits within which they function satisfactorily. There is a lower limit below which they cannot be used to take the measurements. For instance, the ruler/ordinary scale can be used only to take a minimum linear reading of 0.5 mm. If this reading meets the purpose of the testing without affecting the results, the ordinary scale is one of the choices to record the size of the specimen or the displacements the test specimen will exhibit during the loading process. It is the simplest and easiest way to measure the distance between two any given point. Therefore, the level of the significance is one of the most important decisions that the analyst has to take before selecting any device for taking the measurements. If he foresee that a figure of 0.5 mm will not be enough to meet the final objective of the testing as it may result in a significant error in the final results, the only option will be to try some other device that could measure values smaller than 0.5 mm. So, he can try a set of dial gauges, Vernier scale caliper, and micrometer as these have a capacity to take measurements as low as 0.001 mm. If this level of significance is not again serving the intended purpose, the LVDTs can be tried which can measure the deformations to a level of 0.0001 mm or even smaller. This aspect will become still more important when we have to measure the strains the material is undergoing during the loading process. For example, with a least count of 0.5 mm, the ordinary scale can measure a strain of the order of 5000 microstrain only over a gauge length of 100 mm (= 0.5/100). Similarly, if the measurements are taken using the extensometer, this level can be improved to 15–20 microstrain; the magnification of the change in length using an extensometer is usually 1200, therefore they are capable of measuring a minimum length of 0.5/1200 = 0.00041 mm in comparison to 0.5 mm if it is done using the ordinary scale. So, for a gauge length of 25 mm, an extensometer can take the strain measurements up to a level of 16.67 Âμm (= 0.00041/25). The strain gauges are able to measure strains as low as 1 microstrain. However, the use of strain gauges needs to have a dedicated setup, usually the Wheatstone bridge along with a set of duly pasted strain gauges over the test specimen. Any discrepancy in their pasting, alignment or numbers can produce the results in error; the user has to remain very alert during the testing operations.

Similarly, the measurements related to the load applied over the test specimen or the resistance these will offer to the imposed deformations can be taken using the load gauges (just similar to the dial gauges) or the load cells. The adequacy of the load gauge in any test setup depends upon the value of the load increment selected during the loading process or the anticipated value of the material's resistance that would mobilize in the process. For instance, if we want to apply the load in increments of 25 kg till the specimen fails to hold any additional load; the load gauge so selected for this testing assignment must have the least count of 0.25 kN or lower. On the same lines, they must have a rated capacity at least equal to or higher than the anticipated value of the material strength, e.g., for testing 25 mm steel round bars having yield strength of 550 MPa, we require a testing machine with a minimum rated capacity of 400 kN (Check how?). As the test specimen made up of steel usually exhibits strain-hardening during the tension test; so, we have to keep an adequate margin for the possible additional load that the test specimen will take owing to its strain-hardening behavior in the post-yield range while deciding the required rated capacity of the machine or load gauge.

The displacement-controlled testing machines come fitted with all the sensors and a system to impose the deformation in a controlled way to the test specimen. As described in the previous section, these types of the machines are better equipped to measure and record the load–displacement response of the test specimens precisely both in compression as well as tension type of loading conditions. Once the output of the machine in the form of a response curve is known, the material strength (= peak value of the plot), toughness (= area under the response curve), stiffness

(= slope of the tangent at the origin of the curve) of the test specimen can be easily determined. Depending upon the type of the material being examined and tested, the machine parameters (the least count, strain rate and its rated capacity) can be decided. In addition, the users must try to get the following additional information to improve the level of the measurements:

- Accuracy: It is the ability of the machine to tell the true position of its cross-head and is a measure of the maximum error between any two cross-head positions.
- Repeatability: It is the ability of the cross-head to return to the same position when it is moved again and again during the loading process.
- Resolution: It is the smallest mechanical step the cross-head of a machine can make.

The machine having a higher value of all these three parameters should be selected as it allows the user to precisely control the movement of the cross-head during the loading process, which also help to decide the least count of various sensors to be used to measure the deformations (Think, how?).

The type of the test also dictates the machine selection process. Some machines permit to perform the tensile test on specimens, while for the others do only the compression testing. There are other types also where the single machine can perform the testing operations in tension as well as compression loading conditions. Such type of machines is called as the universal testing machines (UTM). The fittings and the grip type used to hold the specimens in all such machines are of prime importance. The maximum force capacity of a grip should match the capacity of the testing machine, which otherwise leads to a premature fracture of the grips itself. A variety of grip types are available; namely, manual vise, wedge, pneumatic, hydraulic type, etc., to hold the test specimens, especially during the tensile loading conditions and each grip can accommodate only certain specimen dimensions based on the opening size of its jaws. However, such type of the constraints are not present when we are evaluating the compression strength of the test specimens and their corresponding response during the loading process (Think, why?). Therefore, depending upon the dimensions (shape/size) of the test specimens and the purpose of the testing: whether tension or compression or shear; a selection of a proper type of the grip plays an important role in getting the true value of the test results. An appropriate selection of grips also helps to avoid the specimen slippage that otherwise occurs during the loading process and leads to errors in the measurements, especially the strains, and elongation, and crack widths and all other derived parameters like toughness, etc.

3.3 Calibration of Equipments/Instruments/Sensors

The testing equipments contain a number of moving components and other accessories in the form of sensors, data acquisition system. During the testing operations, these components move up-down and rotate repeatedly to impose the deformations on the test specimens. In this process; they undergo considerable wear and tear, misalignments, etc. So, the chances are very high that after certain time intervals and/or testing operations, the machines will not be in a position to perform the intended purpose; the rate of the loading/straining may not remain uniform during the entire loading process or it may not be exactly what the user wants to apply. The sensors may report incorrect results. As all these factors have a considerable influence on the strength properties, the malfunctioning of various components if any, needs immediate attention after a certain period of time to avoid possible errors in reporting the results. A process is therefore required to be done at regular intervals to keep the machines in good running conditions and it also helps to keep such type of errors at way. Calibration is one such process that comprises comparing a reading taken on one equipment/device (used in the routine testing operations), with the reading taken on some other machine that has been duly calibrated and referenced to a known set of parameters. It is mandatory that the identical test specimens in these two cases must be evaluated and compared under an identical set of the test conditions. The equipment used as a reference in the calibration process should itself be directly traceable to equipment that has been calibrated according to the standard guidelines framed for this purpose. Normally, all such standard equipments are checked and set as per the guidelines ISO/IEC 17025—International Standard for the accreditation of Testing and Calibration Laboratories.

Calibration thus defines the accuracy and quality of measurements recorded using any testing device or equipment and is a key part of any scientific investigation. It becomes mandatory to carry out such operations when some equipment or device has been repaired or shifted from one location to some other location. It is also required before and/or after a critical measurement or when some equipment, sensor, device has been exposed to a shock or vibrations. Usually, such types of disturbances cause misalignment of various components of the testing equipments that may lead to machine errors in the test results. Any delay to carry out the calibration regularly might be damaging the equipment or leading to some irreparable damage even if we are getting the results within the permitted tolerance range. It is therefore important to keep the eyes open; whenever the observations from some equipment appear questionable or these do not match the expectations from our own experiences or from the output of any other similar equipment, the calibration of equipments must be carried out at the earliest possible, before using all such machines/equipments for any further measurements.

The availability of certified reference material, its traceability and performance verifications of the equipments as per the provisions of the standard reference are the three key check points of a calibration process. Whenever some equipment is purchased, the manufacturer provides a copy of the equipment operation manual. It also contains some information about the various precautions that are required to be observed to safely operate the equipments, etc., along with a suggested set of the periodic maintenance schedule that is needed to be followed for a smooth functioning of the equipments. Ideally, all these listed precautions must be followed while operating the equipments. The maintenance and repair (M&R) operations shall be planned as per the suggested/prescribed schedule for the seamless operations. This type of M&R operations (other than the basic cleaning) is called as preventative maintenance procedures, which keeps the equipments in good running conditions. It should be carried out as per the guidelines of the manufacturers. If such guidelines are not available, these should be framed in-house depending upon the usage of the equipments. The most important thing is that these operations shall be properly documented in the maintenance records, which must contain the following information:

- Type of equipment
- Equipment serial number
- Manufacturer and model
- Location of the equipment
- Date of maintenance
- Adjustments or repairs made
- The identity of the individual performing maintenance and their contact details

In case, the maintenance is performed by an outside vendor, the Incharge of the equipment must retain the original maintenance records provided by the vendor and its entry may be provided in the maintenance record for any future reference.

Like the M&R operations, it is important to ensure that the procedures adopted in calibration process should be appropriate for the intended use of the equipment and shall provide criteria for determining if calibration is satisfactory. The reference material here plays an important role to see whether the calibration process is done satisfactorily or it needs to done afresh. It is vital to check that the reference material/values to be used in the calibration process should be certified by a technically valid procedure or should be traceable to a certificate issued by a certifying organization. In a typical calibration process, equipment to be calibrated is used to take some measurement under the standard test conditions and the output is then compared with the reference material to see the compliance of the test results/observations. In case of any deviation, the necessary adjustments are made in the equipments till the test results starts matching with the reference values. However, if the calibration operations are being carried out by some outside service providers, it is very important to verify their credentials, the competence and the measurement capabilities. The traceability of the reference material going to be used by the service provider in the calibration process should also be verified. The calibration certificates from any such service providers should contain the measurement results, including the measurement uncertainty and/or a statement of the compliance. Similar to the preventative maintenance, the calibration records should be recorded and documented in a register with the following entries:

- Type of equipment
- Equipment serial number
- Manufacturer and model
- Location of the equipment
- Date of calibration

- Certified reference material or reference material used for the calibration
- Copies of all reports, results of calibration and/or certificates of calibration
- A maintenance plan and due date for the next calibration
- The identity of the individual performing the calibration

3.4 Testing Operations

A well-established and the standardized procedures followed to examine the materials experimentally or otherwise with a purpose to determine their key properties is known as testing operations or simply the 'testing.' It forms an important part of the material characterization. Typically, it involved imposing deformations to the test specimen under the applicable test conditions and evaluating their response either in the form of crack pattern, or the resistance being mobilized by them to oppose the imposed deformations, or both. The imposed deformations and the corresponding value of the resistance if plotted on a graph constitute a load–deformation response of the test specimen. Additional information that a testing operation provides is about the cracking pattern and the maximum crack width that the specimen would develop. As described in the previous section, the material strength parameters are controlled by a number of parameters, such as specimen shape and size; type and duration of the loading; type of the equipment and sensors used in the testing operations, a good test engineer therefore must possess an excellent understanding of the working of the equipments/sensors and all possible sources of errors that could be introduced during the testing operations. For instance, the testing specifications for the composites (e.g., Concrete) cannot be same as are used for the metals. Before the testing begins, the test engineer should consider the different available choices of the equipments, the sensors, the dimensions of a test specimen, type of the grips and other measurement devices keeping in mind the level of the accuracy needed to get the desired information and the suitability of each component during the testing.

After deciding the type of testing machine—displacement or load-controlled and sensors, the next most crucial consideration is the position of sensors. Ideally, it should a position that allowed the direct measurement of the desired quantity. A wrong position usually ends up unnecessarily straining of the sensors, and subsequently will influence the repeatability and accuracy of the results. The resolution of the system enhances by a factor equal to the transmission ratio, if the sensor is mounted on the input end of a transmission along with a motor; while its position on the output end of the transmission lead to a reduced resolution albeit with an enhanced accuracy. This aspect should therefore be judicially considered while deciding the possible positions of sensors.

The details of the testing operations needed to evaluate different aspects of materials are standardized by various regulating agencies, such as ASTM, BIS, etc. The test engineer is expected to follow the prescribed testing procedure in totality to get the desired level of the repeatability and the reliability in the test results. The details of various methods and procedure as applicable for different materials are described in the next chapter.

Bibliography

1. Characterization of Composite Materials (2013) Elsevier Science, United States
2. Characterization of Minerals, Metals, and Materials (2019) Springer International Publishing, Germany
3. Frigon NL, Mathews D (1997) Practical guide to experimental design. Wiley, United Kingdom
4. Handbook of Materials Characterization (2018) Springer International Publishing, Germany
5. Jindal UC (2014) Experimental stress analysis. Pearson, Delhi
6. Materials Characterization Using Nondestructive Evaluation Methods (2016) Elsevier Science, Netherlands
7. Practical Materials Characterization (2014) Springer, New York, United States
8. Sensor Technologies for Civil Infrastructures, Volume-1: Sensing Hardware and Data Collection Methods for Performance Assessment (2014) Elsevier Science, Netherlands
9. Sensor Technologies for Civil Infrastructures, Volume-2: Applications in Structural Health Monitoring (2014) Elsevier Science, Netherlands
10. Sensor Technology Handbook (2005) Elsevier Science, Germany
11. Siddique R, Cachim P (2018) Waste and supplementary cementitious materials in concrete: characterisation, properties and applications. Elsevier Science, United Kingdom
12. The Measurement, Instrumentation, and Sensors Handbook (1999) CRC Press, Germany
13. Van der Heijden F, Regtien P, Otthius W, Korsten MJ (2004) Measurement science for engineers. Elsevier Science, United Kingdom
14. Well A, Lorch RF, Myers JL (2010) Research design and statistical analysis. Routledge, United Kingdom

Chapter 4
Material Testing and Evaluation

Having decided the type of materials, the next important task is to freeze their specifications with respect to the procurement/production methods, used as a construction material, the safe stress values and the acceptance criterion for each and every type of the material being finalized. The entire costing of any project depends upon the material specifications adopted in the design of various components and their subsequent construction. For instance, the strength of the material considered in the design of members affects the member sizes and their connections, which in turn controls the construction processes-related cost of the project. In case some low-strength materials have been used in the design, it will require larger member sections, meaning thereby a larger weight of the members to handle (e.g., for the mild steel sections) or more formwork and materials (e.g., for the concrete sections). It will go the opposite way, if some high-strength materials have been employed in the design process. The design process for any project and the related construction process always go side by side; any change in the one part would have a considerable potential to influence the activities in the other part. The starting point in the process, namely materials selection, usually starts from the project site location: First, the materials going to be used in the construction should be locally available and easily procurable; secondly, the manpower required to use and handle them should have the requisite skills and capability; thirdly, the manpower must also be proficient to use the related tools and plants comfortably in the construction activities. Otherwise, the project costing and the construction time will shoot disproportionally as the materials and/or manpower have to source from the far-off locations, which in turn increases the transportation cost and also the time to fetch these things to the site. All such types of points are kept in mind while taking decisions on the material selection/specifications. Once their specifications are frozen, the major task at the site is to ensure that any material going into the construction must meet the identified specifications, especially with regard to their strength, dimensions, handling process, etc. The testing plays an important role to see the compliance of the specifications, in which the information related to the materials going to be used in the project is gathered through various scientific

H. Singh, *Structural Materials*, Materials Horizons: From Nature to Nanomaterials,
https://doi.org/10.1007/978-981-16-3211-2_4

methods and procedures for the evaluation purposes. The data collected during testing can be very useful to the site engineers as well as the design engineers and the project managing officials. The testing is mainly performed for the following listed reasons:

- Selecting appropriate materials and choosing a reliable supplier
- Verifying whether the material received from a supplier is what was ordered
- Meeting the applicable requirements of various regulatory agencies
- Ensuring compliance of the identified specifications
- Forensic analysis to help identify the use of defective and/or inadequate materials, and finally the cause of a failure/rejection if any
- Checking the credentials of new materials before using them in the design process.

Testing and evaluation help the engineer to understand and quantify whether the materials selected for a certain project are suitable for the use or not. The structural stability and safety of any building or part thereof depend entirely upon the quality of the materials, both its raw (such as cement, sand, aggregates, bitumen, steel rebars, etc.) and the finished form (concrete, asphaltic concrete, etc.), going into the construction. The improper material selection or non-compliance of the identified acceptance criterion during the execution of the project may lead to serviceability failure where the components fail to serve its intended function. The discrepancies also lead to multiply the avoidable maintenance and repair costs of the project. This chapter presents various steps that are usually followed to ensure compliance of the specifications along with different methods to evaluate various material characteristics.

4.1 Sampling

Almost all materials differ in the response despite their identical composition and production processes. The variations exist even in different specimens prepared from the same material. Because of these variations, it becomes a big challenge for engineers to ensure consistency in the material properties. It becomes mandatory for them to get the strength properties at least equal to that taken in the design of various components of the building, etc. In case of large variations, the materials have to be rejected or members need to be redesigned for the actual value of material properties they are getting on the site. For instance, if some member has been designed assuming M40 grade concrete (that indicates its 28-day compressive strength obtained from the standard cubes), the site engineers are supposed to get a compressive strength of at least 40 MPa (or a value as per the acceptance criterion; it may be 45 MPa or even higher depending upon the permitted standard deviation) to use that lot of the concrete. If they are unable to do so, because of any reason, such as unskilled workforce, inadequate tools and plants, and non-availability of good quality raw materials in the nearby areas, the only option for them will be to get the structural design reviewed and redesign the members using some lower concrete grade (maybe M25, M30 or so) that they can produce locally without compromising on the quality of the end product, else arrange the materials having the desirable quality even from the distant locations/places. Either way, the project construction cost is going to escalate, in the form of either a higher material consumption (the first case) or the higher transportation cost and the associated time to fetch the materials (second case). Engineers have to judicially take the final call in the overall interest of the project.

Generally, a large quantity of materials goes into the construction. Ideally, the site engineers need to test the entire batches of the materials going to be used at the site. Such an approach will be very time consuming, and also, it leads to increase in the construction cost caused by the extra material being required to prepare the test specimens and the overhead cost associated with various testing operations. Different statistical methods come handy to help engineers to optimally manage the show and ensure that they are getting the material properties what they are supposed to. These methods follow structured approach. They try to ensure compliance of the specifications by means of sampling—a process used in statistical analysis in which a predetermined number of observations are taken from a larger population being examined. A subset constituting a finite number of the observations taken as a representative of the population is called as a sample, whereas everything in its entirety constitutes a population whose properties we wish to investigate or study through the testing. For example, while assessing the overall quality of the concrete being used in the construction of some building project, the total concrete being used in the construction is the population, whereas a few number of the cubes filled from various members (say, a set of 3 cubes from each) of the building constitute a sample from the population. The site engineers try to get the desired information related to the materials being used in the construction by testing the samples. For the statistical analysis, it is important that various test data points, called as attributes, to be considered in the sample should be continuously measurable quantities, such as material strength and physical dimensions of various test specimens. The outcome of the statistical analysis forms the basis to determine the overall acceptability of the batch of materials used or being used in the construction. There are a number of sampling methods, described below:

The sampling process can be broadly categorized as the probability-based and non-probability-based sampling. In the former case, every unit in the population has a probability (greater than zero) of being selected in the sample, and this probability can be accurately determined. Whereas in the non-probability sampling, some units of the population have no chance of their selection or the probability of their selection cannot be accurately determined; so, it normally leads to bias in the process thereby putting limits on the extent of the information that can reliably be obtained from such samples.

The probability-based sampling has further classes formed on the basis of selection procedure used in the process. It can be a simple random sample (SRS) type where all the subsets taken randomly from the population have an equal probability of being selected. This type of the selection minimizes bias and simplifies the process. However, it is vulnerable to the sampling errors because of the randomness of the selection process that may produce a sample that does not reflect the makeup of the entire population. Moreover, this process is cumbersome and tedious when the sampling is to be done from a large target population. For example, a simple random sample of some forty students from a college will on average produce twenty boys and twenty girls, but any given trial during the sampling is likely to overrepresent one sex, while under-representing the other as they are sampled randomly without seeing their characteristics. This type of the flaw is removed in the other two forms of the sampling process, which attempt to overcome this problem by 'using the information about the population' to choose a more 'representative' sample instead of some blanket selection. These two techniques are known as systematic sampling and stratified sampling.

In the systematic sampling, the sampling points are picked up at regular intervals from the population pre-arranged in some defined sequence. For instance, the students in the above-mentioned example can be lined up in ascending order (maybe, on the basis of their height, etc.) and then every third student (say) is recorded for the sampling purposes. It can be started from any student randomly and then proceeds with the selection of every third student. It is important to note that the starting point in the process is not automatically the first in the list, but is instead randomly chosen from within the first to the third student in the list. On the other hand, a simple random selection could easily end up with too many boys or girls, leading to an unrepresentative sample. However, selecting every third student ensures that the sample is spread evenly across all the students considered in the population, thereby eliminating the bias in the selection. However, this type of sampling is vulnerable to periodicities in the list. If the periodicity is present and the period is a multiple or factor of the interval used to pick the sample, it is likely to be unrepresentative of the overall population, making the technique less accurate than the simple random sampling.

For instance while constructing RC buildings, a number of cubes are filled on-site to check the quality of the concrete being poured during the construction. These cubes are filled from the different batches of the concrete produced to cast different members of the building, like footings, columns, beams, slabs, etc. Depending upon the quality of the concrete and the placement and compaction operations adopted at the site, the weight of different hardened concrete cubes varies specimen to specimen. Many times, the surface finish of some of them, especially their top face, will also be uneven. As the strength of concrete cubes resides on the solid part, variations in the weight indicate a change in the microstructure of the concrete; the heavier cubes will have a dense microstructure and consequentially will exhibit a higher strength. The uneven top surface or the sides will also affect their load carrying capacity.

Say from a population of 200 cubes, we are picking any 20 cubes for testing purpose. In the simple random sampling, these 20 cubes will be picked up from the entire population of 200 randomly, irrespective of their weight, unevenness or other features. We can expect these 20 cubes from the best of the lot, all having the maximum weight, no uneven surface; or the other way round, a sample of cubes having the lowest weight; or a mix of the dense and weak cubes, some with a level surface while others with an uneven surface. A lot of bias will be present in the sample, which will affect the final results.

On the other hand, if we arrange all the cubes in the entire population in ascending order of their weights and pick every fourth cube (say) up from the line for the sampling purpose, it will reduce the bias in the results (it will not be zero!). It is important to recall that the first pick can be any cube in the entire lineup, and not necessarily the first cube is the first pick in this systematic sampling. However, there again exists a possibility of sampling error as the cubes having a heavier weight in the lineup may have uneven top surface or other irregularities, or vice versa may be also true. As these types of irregularities are of a local nature, their influence is only limited to the cube strength and is not going to affect the strength of the members (being) cast from this lot of the concrete. However, it will definitely affect the results we are expecting from the cubes and cannot be considered a true representative of the concrete used in the construction.

This type of the biasing and errors in the sampling can be removed to a significant extent by shifting to the third type known as a stratified sampling approach. The overall procedure in this approach is almost identical to that used in the systematic sampling, but there is a slight difference in the approach used to pick up the specimens. The cubes are selected from a pre-decided intervals (say, every fourth), but it has to ensure that 'the pick' should have a level surface (or without having any other irregularity that can locally influence the cube strength). For instance, if the eighth cube is not evenly cast, then 'the pick' can be seventh or ninth cube without having any irregularity. Then, the next 'pick' will be the fourth from the last 'pick' position (e.g., 11th or 13th). So, the sample will have cubes that are fit for the testing to see the quality of concrete used to pour the respective members, the effect of the compaction and placement operations used in the casting process, etc.; such a sample can be considered as a true representative of the concrete used at the site. The different strata in the approach can be various members of the building; e.g., the cubes from the footing can be used to assess the quality of concrete used in the casting of the footings. Likewise, it can be for the columns, the beams and so on. This way, the bias in the sampling is removed by a 'selective pick' instead of the mechanized nature of the approach adopted in the simple random sampling or the systematic sampling approach. This approach is more refined and scientific, but is more time consuming as it will take relatively more time to select the sample.

The non-probability-based approaches, such as accidental sampling and voluntary sampling should never be used to quantify the level of material quality. These two approaches try to get the samples from that part of the population, which is readily available and accessible. It may be done through meeting the person or including a person in the sample when one meets them or chosen by finding them through technological means such as the Internet or through the phone. For instance, the cubes cast from the concrete used in the footings cannot be used to determine the level of concrete quality used in the columns or beams; the cubes have to be taken from the concreting operations used to cast the footings only,

Table 4.1 Minimum number of samples for concreting work

Quantity of concrete, m^3	1–5	6–15	16–30	31–50	>50
Number of samples	1	2	3	4	4 + one for every additional 50 m^3 or part thereof

Source https://archive.org/details/gov.in.is.456.2000

meaning these should be relevant one. The test data generated by using the samples collected from the non-probability-based approaches cannot make scientific generalizations about the entire population, being not a true representative of the population.

It is important to keep in mind that an attempt should be made to avoid the selection bias in the sample; it generally occurs when the probability of the true selection of a specimen differs from those assumed in calculating the results. This means if we have taken equal probability of the occurrence, then the sampling should be done for the entire lot under considerations. Neither there should be an overcoverage (inclusion of some data points from outside of the population), nor any under-coverage in the selection of a sample from the population. Once the sample is appropriately selected, it is examined to fetch the desired information based upon the experimental studies or otherwise. Different guidelines prescribe procedures to take samples depending upon the materials; e.g., IS 1199 gives the sampling procedure for the fresh concrete. It states that at least one sample shall be taken from each shift. Where concrete is produced at any continuous production unit, such as ready-mixed concrete plant, the frequency of the sampling may be mutually agreed upon by the suppliers and the purchasers. Table 4.1 lists the number of samples (each having a minimum of three cubes) as per the provisions of IS 456 that must be taken depending upon the quantity of the concrete being poured.

The cubes in the samples so finalized are tested under the standard conditions to know their compressive strength (maybe 28 days). The test data form the basis of acceptance or rejection of the concrete quality.

4.2 Acceptance Criterion

At the macro-level though some materials look homogeneous, most of them are inherently heterogeneous. It leads to the variation in the response of different specimens prepared from the same lot of the material, even under a similar set of the conditions. Normally, the changes in the material microstructure caused during their production are responsible for the scatter observed in the response. Dimensional variations also exist that lead to the change in the load carrying capacity of the members. In such a scenario, it becomes obligatory on the part of the engineers or the officials responsible for the quality control at sites to sample test specimens, appropriately, to get a truly representative sample for the testing and evaluation purpose.

If large variations exist in the test results obtained from the sample, the use of such material is definitely going to affect the safety and integrity of the members/buildings. Some lower and upper limit, therefore, has to be imposed on the quality of the material to get the satisfactory performance during the lifetime of the member/building. Any such limits have to be properly documented before the start of a project, and it should be mutually agreed by all concerned stakeholders. The pre-established standards and/or all documented requirements of a project that any material or the products made from such materials must meet to perform satisfactorily during its intended design life are known as an acceptance criterion. Each material requires a different set of acceptance criterion and these are usually prescribed by various agencies in the form of the design codes and/or guidelines. The quality of the materials or products made from such materials and their associated cost is directly linked to the acceptance criterion adopted to define the level of the quality control. An attempt should be made to identify such guidelines and follow these in the true spirit to get the desirable standard of the quality defined through them. In the absence of suitable criterion, or if the purpose is to get something better that is offered by the available guidelines, acceptance criterion can be framed very specific to the needs of the project in consultation with all the stockholders. Following points must be kept in mind while framing any acceptance criterion:

- It should be written before implementation and should be independently testable.
- It should be compatible with the design philosophy used to proportion different members/components in the project.
- It should be very clear and concise, without any ambiguity. Acceptance criteria must have a clear pass or fail result; i.e., the material under considerations is either accepted or rejected based upon the test results. There should be no room for any interpretation.
- Every stockholder of the project must understand and agree on the criteria, and it should be a part of the work order/contract document.
- It should focus only on the end result and not on the solution approach used to achieve the results.

It is worth to recall that the material test data and other physical dimensions of interlocking tiles, bricks, concrete blocks, reinforcing steel bars and other building materials generally used in civil engineering projects follow a normal distribution. Figure 4.1 shows a typical plot of this type of the distribution. It indicates that the data near its mean value are more frequent in occurrence than the data far from the mean value; around 68.26% of the data points fall within an area enclosed between the limits set by the standard deviation of $\pm 1\sigma$. This area increases to 95.44% if the corresponding limits are shifted to $\pm 2\sigma$. This feature of the data distribution helps to formulate acceptance limits and is being conveniently employed in practice to ensure the desired level of the quality control at the construction sites.

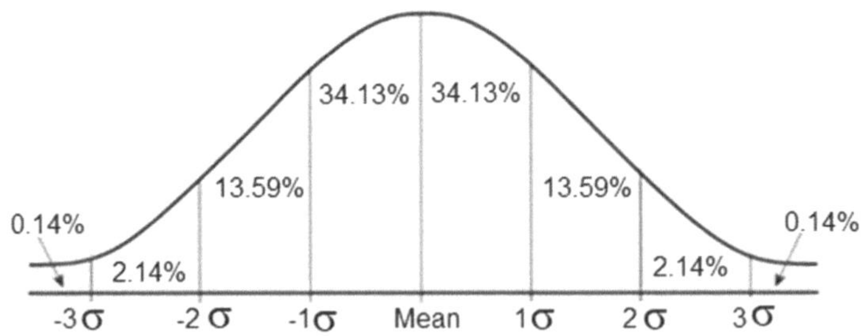

Fig. 4.1 A typical set of data that follows a normal distribution, showing the percentage of the area that falls within the limits defined through the standard deviation of data. *Source* Author

By appropriately defining the upper and lower limits of the standard deviation, an acceptable level of the batch quality can be set as a target to achieve. Within these limits, an acceptable quality level can be defined as a maximum allowable fraction of the defective attributes. With an upper limit of $+\sigma$, the fraction of the non-desirable values in the data set is equal to the area under the curve to the right of this limit. This value is then compared to the allowable value to determine the acceptability of the batch. Similarly, the values on the left of the lower limit can be determined. Alternatively, the limits (both, upper and lower) can be decided by starting from any desirable value of the sample mean value. Using this procedure, we can easily set a target of achieving that value of the desirable probability of occurrence of the value of the sample mean. Characteristic value of the attributes (maybe strength, dimensions, weight of test specimens, etc.) is usually defined to achieve this objective. For example, the characteristic strength (f_{ck}) of the material is usually adopted in the design of any structural element instead of an average or some other arbitrary value of the data; it is defined as the strength of material below which not more than 5% of the test results are expected to fall. This means that the mean value of the material strength has to be significantly greater than the 5 percentile characteristic strength considered in the design of structural members. Physically, strength is a measure of the resistance that any material provides to the imposed load, below which the chances of its failure are only 5%. For example, concrete having a characteristic strength of 25 MPa means if a stress of 25 MPa or less is developed in any given concrete member, the probability of its failure is only 5%. This way of expressing the material strength forms the backbone of the reliability-based structural design methods. We can easily achieve this target if we adopt the lower value of the acceptable limits for the material test data from its mean strength value as f_{ck} = sample mean—1.64σ, where the parameter (σ) is the standard deviation of the sample being examined and f_{ck} is the characteristic strength of the material. On the other hand, if the characteristic strength (f_{ck}) of the material is specified, then the sample must possess a target strength (f_t) computed as $f_{ck} + 1.64\sigma$. As is indicated in Fig. 4.1, if we take the sample standard deviation of 2, instead of 1.64 in these expressions, the probability of failure will reduce to 2.14% only.

Sometimes, a single acceptance criterion is not enough to decide the acceptance of the batch of any given material, especially when the response of the material is not consistent under different loading conditions. For instance, concrete responds differently to the compressive stresses and the flexural stress, usually very strong under the compressive loading, while it exhibits a brittle behavior under the flexural loading conditions. In such a scenario, a set of the criterion has to be defined to see the compliance and to ensure a satisfactory performance of the concrete under different stress conditions; it will be different for the different strength parameters being evaluated. For instance, the design code IS 456 has prescribed two separate criterions for the compressive strength and the flexural strength of concrete. Both are based upon the characteristic strength (f_{ck}) of the concrete considered in the design of structural members; these limits are given in Table 4.2.

Normally, acceptance limits based upon some alternative way of testing (non-destructive or semi-destructive methods) are also prescribed by various guidelines to ensure achievement of a certain minimum level of the desired concrete quality, if the test data from the sampled cube (destructive testing strategy) fail to give a clear-cut decision on the acceptance. For all such types of the cases, IS 456 prescribes testing of the cores obtained from the hardened concrete. The prescribed acceptance criterion states that 'concrete in the member represented by a core test shall be considered acceptable if the average equivalent cube strength of the cores is equal to at least 85% of the cube strength of the grade of concrete specified for the corresponding age and no individual core has strength less than 75%.'

Acceptance limits also need to be finalized when the value of some parameter is vital to the safety of the component or the building as a whole. For instance, the position of the reinforcing bars defined through the cover requirement along with the dimensions of the reinforced concrete members is very important to ensure achievement of the required design flexural capacity. A large deviations from the values considered in the design will lead to a reduction of the flexural capacity and other unexpected behavior. IS456 set an acceptance limit of +10 mm/0 mm on the actual value of the cover to be used in the member from its required nominal cover. Similarly, it is +12 mm/−6 mm on the dimensions of RC member and ±10 mm and ±15 mm in the position of rebars placed in members having an effective depth up to 200 mm and more than 200 mm, respectively. Same way, the limits have been put on the maximum variations in the diameter and weight of the reinforcing

Table 4.2 Acceptance criterion for concreting work

Mean flexural strength (MPa) computed from		Mean compressive strength (MPa) computed from	
Any single specimen	A group of 4 consecutive specimens*	Any single specimen	A group of 4 consecutive specimens*
$\geq f_{ck} - 0.3$	$\geq f_{ck} + 0.3$	$\geq f_{ck} - 4$	$\geq$ Higher of ($f_{ck} + 0.825\sigma$) or ($f_{ck} + 4$)

*σ is the established value of the standard deviation and it must be computed from the test data of 30 cubes

Source https://archive.org/details/gov.in.is.456.2000

bars; these are ±0.5% (up to 25-mm diameter), ±0.6% (25–35-mm diameter), ±0.8% (36–50-mm diameter) and ±7% (up to 10-mm diameter), ±5% (10–16-mm diameter), ±3% (>16-mm diameter), respectively. The purpose of imposing these limits is to just ensure that the member to be cast using the respective material, placed within the permitted range of their tolerances, must be able to develop the required design flexural capacity. In case the variations observed during the routine testing are found to be more than their permitted values, either the materials should be rejected or the concerned members must be redesigned/recast considering the actual values of the parameters in the design/construction process, respectively.

Consider some other example to visualize the effect of a poor acceptance criterion on the costing of a project. In the brick masonry construction, bricks and cement mortar are the two major materials. Ideally, all bricks to be used in the construction work should have a uniform size; generally, the brick size 225 mm × 115 mm 75 mm is prevalent in most part of India. If we are constructing a masonry wall using the bricks having huge variations in their sizes and the adopted acceptance criterion is permitting their use in the construction (say, for the purpose of illustration). In such a case, it is not possible to get a uniform wall thickness. Some bricks will protrude out from the wall surface, thereby giving an uneven surface finish. Assuming an average brick size of 225 mm with a maximum size of 238 mm and a minimum of 215 mm in the lot, the maximum undulations in the finish surface will be around 23 mm. Therefore, in order to get an even surface finish, we need to plaster the wall with a thickness of 23–25 mm against a normal requirement of 12 mm! It will unnecessarily increase the cost of the materials and the man-hour needed to complete the job. On the other hand, if the acceptance criterion is not allowing the use of this lot of bricks, the quality of the work will improve significantly with additional economic benefits. But in that case, we have to procure the bricks that meet the acceptance criterion to ensure achievement of this level of the perfection.

Appropriate sampling of the materials and the acceptance criterion are thus two main pillars of any testing and evaluation process. The sample of materials used in the construction work or to be used in the work is tested as per the standard guidelines and the test data are evaluated as per the acceptance criterion adopted for the said site/project. Usually, the criterion framed by various design guidelines is followed because these are framed after due deliberations and evaluating the effect of variations in the material/member/product dimensions, their weight and strength on the design capacity of members and overall economics of the project/work. Sometimes, the prescribed standard acceptance criterion is also customized to meet any special requirements of the project/work. The final costing of any project and its overall quality is a direct outcome of the acceptance criterion adopted at the start of the project. It is therefore a standard practice to freeze the acceptance criterion and the evaluation procedure at the time of awarding the contract to any construction firm/agency. The adopted acceptance criterion should be made very clear to all stakeholders of the project to avoid any litigation or delay in its timely completion.

Example: M20 concrete is used in the design of RC building. The test results pertaining to 28-day compressive concrete strength (in MPa) are 24.89, 30.22, 25.78 and 26.22. We have to check the acceptability of the concrete?

The acceptance criterion given in Table 4.2 can be used to see whether the concrete is acceptable for use or it should be rejected. It states that any concrete sample is acceptable if the individual reported strength is coming more than its characteristic strength minus four and the group results must meet the other requirement listed in the table. The summary of this exercise is tabulated below:

Cube no.	Compressive strength (MPa)	Individual data point (MPa)	For a group of four samples, higher of the following two values (MPa)	
		$f_{ck} - 4$	$f_{ck} + 4$	$f_{ck} + 0.825\sigma$
1	24.89	16.0	24.0	21.95
2	30.22	16.0	24.0	21.95
3	25.78	16.0	24.0	21.95
4	26.22	16.0	24.0	21.95
			Higher of the two parameters = 24.0	

Source Author

The average compressive of the sample is computed as 26.78 MPa. It is important to note that the sample must be of a minimum of 30 specimens for using the test data from it to compute the standard deviation. In the present case, it is worked out from the four data points as 2.36 MPa. The first part of the criterion says that all individual value must be more than the 16 MPa; it meets the condition. The second part of the criterion also meets the listed condition as the sample average (=26.78 MPa) is more than the listed requirement (>24 MPa). As such, this concrete sample is acceptable for use for its intended purpose. If we calculate the actual value of the characteristic strength, it is obtained as 22.9 MPa ($=26.78 - 1.64 \times 2.36$); all the individual cube strength values are more than this value; therefore, we can use this concrete lot confidently for the intended construction activity. It meets the definition of the characteristic strength (check, how?).

4.3 Different Test Parameters

The very purpose of structural materials is to help engineers transfer safely the load from the point of their application to any other point of interest in members. During the load transfer process, the materials will deform and undergo some change in their shape and size. These deformations help them mobilize the inherent material's resistance to counter the applied stresses caused by the external loads. In this way, all materials to the extent possible, try to maintain equilibrium between the internal and the external stresses in the structural members without losing their integrity and maintaining their pre-assigned shape and size. They will give up only when the materials are used to make them exhaust their capacities. For example, any material can be loaded and unloaded a number of times within their elastic limits and the member will never crack or deform permanently up to this limit. As soon as the load effect increased beyond the elastic limit, the materials will start using their reserve strength lying in the post-yield range, albeit it happens with minor to very severe cracking and deformations depending upon the type of material used in the construction and the magnitude of the applied load.

It is, however, always possible to design the members to perform extremely well in their post-yield range. The material characteristics, either known from an established data or determined by means of different testing operations, play a key role in the design process. Usually, the following material characteristics are needed to proportion and design any structural member to perform their intended functions.

4.3.1 Strength and Stiffness

All materials deform under the action of external load; it can be a thermal loading or some mechanical loading. The extent of the deformations depends upon the type of material being loaded; the brittle materials fail suddenly, while the ductile ones go on resisting the applied load and during this process, they usually undergo large deformations before they finally fail. They exhibit a load–deformation response pattern very unique to each and every type of the material being tested. The maximum value safely supported by them before the failure, the corresponding value of the deformations (both along the loading axis and normal to it) and the load value corresponding to the onset of the yielding and rupture/crushing are the key observable points during the material testing operations. The maximum value of the load sustained by the materials during the loading process determines their strength and it can be defined as their ability to sustain force without any form of failure; the stronger the material, the greater the force it can withstand. All materials find their structural applications only because of this property as it determines their capacity to support the applied load and safely transfer it to some other point of interest in the structural member. However, the way any material responds to the external loading is also equally important in the design process. The load–deflection curve obtained during the testing describes the response of materials to the applied load. The initial slope of this curve defines the material stiffness, k ($=\Delta p/\Delta$), and it can be calculated by dividing the load increments (Δp) to their corresponding change in the deformations (Δ) exhibited by the material sample under the effect of applied load. In other words, the stiffness (k) of a material is a measure of the magnitude of the force needed to deform the specimen by unity. A high value of the stiffness indicates that the material under investigations is very stiff and it will need a large magnitude of force to deform it and vice versa. These two material parameters are the key to any successful design: The former fixes the load carrying capacity of the member, while the latter determines how it would respond to the applied loading and undergo deformations. As discussed in the previous chapters, the internal microstructure of materials plays an important role in deciding their strength and stiffness. Normally, a dense packing of the constituent particles in the material makes it stronger than the material having a porous microstructure; it happens because of the stronger internal bonding between the building blocks of the materials when these are closely packed.

Strength is an intensive material property and is calculated as a ratio of the maximum load (P) carried by the specimen just before its failure to its cross-sectional area (A)—always, taken normal to the loading axis ($=P/A$). It is expressed as a ratio of the force units (N, kg, kN) and the area units (mm^2, cm^2, m^2), e.g., N/mm^2 (=1 MPa), whereas the stiffness is expressed as a ratio of the force unit (N, kg, kN) and the deformation units (mm, cm, m), e.g., N/mm. Each material possesses a unique strength (σ) and stiffness (k) value, albeit it will be different if we test the same material specimen under a tension load or a compression load or shear load or the bending action. For instance, the plain concrete is very strong if it is tested under a compression type of loading but its load carrying capacity is reduced to about one-tenth of its compressive strength if the nature of the loading changes to the tension. Though the material in this case remains the same, its response changes with the change of nature of the loading; the compression type of the force brings the particles closer, while these move away from each other under a tension force thereby affecting the material response, which entirely depends upon the extent to which different stress transfer mechanisms are mobilized in the material. Similarly, the plain concrete has very low value of the bending resistance; it suddenly fractures into two segments once loaded under a 3-point loading, etc. (check out the possible reasons?). However, reinforcing the concrete either internally or externally greatly modifies the inherent brittle characteristics of the same specimen and makes them to behave like a ductile one.

Example: The maximum permissible displacement for a concrete cube of 150 mm size is 2 mm. What should be the highest magnitude of vertical load that we can apply to the concrete block? The stiffness of the concrete used in the production of the block is 23,500 N/mm. And, check whether the concrete specimen if subjected to that much loading will crush or not? The compression strength of the concrete used to cast the cube is 20 MPa.

It is given that the concrete specimen should deform (Δ) by an amount of not more than 2 mm. The stiffness (k) of the material is also given as 23,500 N/mm.

By definition, if the specimen is loaded in its elastic range (say), the stiffness $k = \frac{P}{\Delta}$ can be used to determine the load magnitude (P) that compresses the specimen by 2 mm.

$P = k\,\Delta = 22{,}500 \times 2 = 45{,}000\ \text{N} = 45\ \text{kN}$.

This means that if we apply a load of 45 kN over the face of the cube, it should deform by 2 mm and will stress the specimen by a magnitude of P/A, where A is the cross section of the test specimen (=150 × 150 mm). If the stress value so applied exceeds the strength of the material, the chances are that it would crush.

Therefore, the maximum value of the stress in the specimen = $\frac{45{,}000}{150 \times 150} = 2\,\text{N/mm}^2$

This means that the specimen will not crush as the value of the stress caused by the applied load of 45 kN is very small in comparison with the stress value that the cube can sustain in its elastic range. Usually, it is equal to one-third of its compression strength (=20/3 = 6.67 MPa).

The other way round, if the strength of any material is known (from some experimental test data, etc.), then we can very easily determine the cross-sectional area of an object needed to support the given value of the axial load. For example, the design compression strength of brick masonry is of 0.35 MPa and we have to place a load (*P*) of 75 kN/m (say) over its entire length. From this set of data, the plan area of the brick wall needed to support this load can be easily calculated as $P/\sigma = 75{,}000/0.35 = 214{,}285\ \text{mm}^2$ for every one meter of the wall length. This area (=thickness × length) can be provided by suitably placing the brick in the cement mortar; e.g., the wall thickness (*t*) can be calculated as $A/1000 = 214{,}285/1000 = 215$ mm (by taking 1000 mm long stretch of the wall from its given length). So, we can achieve the required strength by using one brick thick wall, which will provide an area of $225 \times 1000 = 225{,}000\ \text{mm}^2$ ($>214{,}285\ \text{mm}^2$) for every one meter of the wall. Providing an area more than that required from the strength consideration is always safe, albeit it would increase the cost of construction if provided in excess (think, why?).

Knowing the stiffness of the masonry (=25 kN/mm say), the resultant deformations that the wall will undergo under the applied stress can be computed as follows:

$$k = \frac{P}{\Delta} = \frac{\sigma A}{\Delta}$$

$\Delta = \frac{\sigma A}{k} = \frac{0.333 \times 225{,}000}{25{,}000} = 3.0\text{mm}$, where σ is the actual value of the stress being produced in the wall section (=75,000/225,000 = 0.333 MPa).

However, the value of the deformation that the masonry undergoes on application of the load can be decreased by reducing the stress level in the wall section. The smaller the stress level, the lesser will be the resultant deformation, as is indicated in the above equation.

Another important application of this concept is the estimation of a safe stress level that any material can withstand; e.g., a soil mass generally fails in shear at the ultimate load level, but during the loading process, it goes on deforming and this process can be observed from the settlements that a rigid wooden block (say) placed over the soil surface exhibits under the increasing load levels. Let us assume that the soil mass fails to support any further loading, when it is loaded beyond a stress level of 230 kN/m^2. However, the soil mass will be deformed by a certain amount at this stress level (=85 mm, say). If the building is capable or it can be designed (say) to absorb this much deformation, the footings can be proportioned using a soil pressure of 230 kN/m^2; otherwise, we have to pick some other value of the stress from the soil load–deformation response curve that the building can withstand safely. For instance, if the soil mass deforms by 25 mm (=permitted value of the footing settlements) corresponding to a pressure of 90 kN/m^2, in such case, the safe soil pressure for the design of footings will be 90 kN/m^2 (=smaller of 230 and 90 kN/m^2), so that the footings should deform by not more than 25 mm at the design load. This way, there exist a number of options to transfer the given load through the material. Engineers have to select the most suitable stress level and a suitable shape/size of the body/member that meet their design constraints.

4.3.2 Toughness and Resilience

As said in the previous section, all solids deform during their loading. During this process, they go on absorbing and storing the energy supplied by the external loading. These two concepts, namely toughness and resilience, measure the amount of the energy that a material can store just before its failure and up to their elastic limit, respectively. The more energy they can store safely, the higher will be their capacity to withstand the applied loads and the more will be deformations that they will be able to undergo before their failure. This type of the material response is quantified by a strength parameter called as toughness. By definition, it is the ability of materials to deform, plastically, without fracturing.

When certain stress (σ) is applied (gradually) to some material, it produces a set of the deformations during the process. These can be along the axis of the applied loading as well as along the axis transverse to it. The resultant deformations along the loading axis (say) are expressed in the form of strain (ε). A small amount of the work (δW) performed by the imposed stress (σ) in producing a strain increment of $\delta\epsilon$ during the loading process can be determined from Eq. 4.1.

$$\delta W = \frac{1}{2}\sigma\delta\varepsilon \tag{4.1}$$

The total work done (W) by the imposed stress during the entire loading process can be found by integrating the work function given in Eq. 4.1. The resultant expression will indicate the sum total of the area under the load–deformation response curve of the material (check, how?). The SI unit of the toughness is Joule per cubic meter (J/m^3). In order to be tough, therefore a material must be both strong and ductile. For instance, the brittle materials (like, the plain concrete) are strong, but they have a limited ductility; conversely, very ductile materials (like, the rubber) with low strengths are also not tough. Therefore, for a material to be tough, it must have a capability to withstand both high stresses and high strains.

It is important to recall that the strength indicates how much force the material can support, while the toughness indicates how much energy a material can absorb just before its failure. The internal microstructure of the material plays an important role in deciding the toughness. The most important part for a material to be tough is that it must be able to continuously switch the five major stress transfer mechanisms, listed in Chap. 1, from one to another without fail; the easier it can do, the tougher the material will be. Out of all the materials, the composites are the only type that can be designed to have a high strength as well as the desired value of the (high) toughness, just by playing with their microstructure. For example, by mixing an appropriate dose of short steel fibers in the concrete during its production stage make it stronger as well as tough. The reason is simple: The plain concrete is brittle in nature, so as and when the applied load exceeds its threshold value the concrete fractures. However, the (uniform) presence of steel fibers in the concrete will not allow it to respond that way. After the loss of the internal bonding between different constituent units/particles within the concrete mass, it continues to support the load by means of the internal friction and the mechanical interlocking being mobilized along the uneven fractured surfaces in the member along the slip/failure planes and, subsequently, the bridging action of all those fibers which are crossing the fractured surfaces play their role in making it tough. Generally, all structural components that are required to support a heavy static load as well as an impact loading are designed to be strong as well as tough and the composites (e.g., reinforced materials) are the most suitable materials for this purpose.

On the other hand, the resilience is the ability of a material or objects made from it to spring back, elastically, into its initial shape/form. It is a measure of the capacity of any material to absorb energy elastically, whereas the toughness is the total energy absorbed by the material just before its failure (it includes elastic as well as the plastic range of the load–deformation curves). Therefore, by considering the area of the load–deformation response of a material up to its elastic range, the resilience can be calculated. Springs, shock absorbers, etc., are usually designed on the basis of the material resilience because the purpose of their use is to help the supported members (by them) to bounce back quickly to their original form, quickly and repeatedly, without showing any sign of distress/failure. It is important to note that the chance of any permanent damage to either the supported member or the spring will be low as they are designed to operate within their elastic limits on the basis of material resilience.

Example: Two standard fiber-reinforced concrete beams are tested under the 3-point loading conditions. Typical data for some specific beam specimens are obtained from the UTM in the form of load–displacement response; it is given below:

Displacement (mm)	Load (kN)	
	Sample-1	Sample-2
0	0	0
0.5	15.5	15.5
1.5	12.2	19.25
2.5	9.3	21.25
3.5	7.4	23.05

Source Author

Calculate the material resilience, toughness and modulus value from these data. Decide the type of the material from this response data (i.e., strain-hardened, strain-softened, elastic, plastic, elasto-plastic, etc.).

The data given for the sample-1 can be converted to the load–displacement plot easily using the spreadsheet. It is shown below:

The curve in the plot indicates that the said material sample is softening with increasing load. It can be easily judged by seeing the post-peak point of the curve, wherein the load carrying capacity of the specimen reduces with the increasing load. The maximum load sustained by the material (=15.5 kN) represents the material strength; it can be converted to 'MPa' units by applying the bending formula, if required. It is convention to express the strength of any material in the form of a stress at the peak load and is given in MPa or kg/cm^2 or other similar units. On the other hand, the toughness of the material is indicated by the complete area under the curve, right from the origin to the failure point. It can be calculated by counting the number of small boxes under the curve; the area of each box is 0.2 kN mm (=2 × 0.5/5). Alternatively, it can be computed by splitting the area between the curve and x-axis to multiple trapezoids, then calculating the area of every trapezoid individually and finally summing up these areas. It is tabulated below for the sample-1:

Displacement (mm)	Load (kN)	Material toughness (= Area under the load-displacement response curve) (J)	Cumulative sum of values in 3rd column (J)	Material property, Joules (=kN mm)
0	0	3.875	3.875	
0.5	15.5	13.85	17.725	Resilience
1.5	12.2	10.75	28.475	
2.5	9.3	8.35	36.825	Toughness
3.5	7.4	–		

Source Author

The tabulated data show that the toughness of the given material in the case of the sample-1 is around twice the value of its resilience (=area up to the LoP, taken approximately equal to its peak load). The elastic modulus/stiffness of the material can be computed by measuring the slope of the tangent drawn at the origin of the curve. It is given below:

$$= \frac{14 - 0}{0.4 - 0} = 35\text{kN/mm}$$

On the similar lines, the slope of the curve in the post-peak region can be determined; it is given below:

$$= \frac{10 - 14}{2.3 - 1.15} = -3.48\text{kN/mm}$$

The negative slope in the post-peak range is an indication of the material softening response; it indicates the loss of the load carrying capacity that the beam sample-1 would exhibit due to cessation of one or more stress transfer mechanisms being mobilized in the material. The slope should be appropriately computed if the curve is depicting continuous changing slope.

If we repeat the same procedure for the sample-2, the load–displacement response curve can be plotted and is given below:

Unlike the beam sample-1, the rise of the curve in the material post-peak range indicates that the specimen is exhibiting the material hardening response. The data for the sample-2 shown in the curve can be processed to find the material properties. It is tabulated below for a ready reference.

Displacement (mm)	Load (kN)	Material toughness (=Area under the load-displacement response curve) (J)	Cumulative sum of values in 3rd column (J)	Material property, Joules (=kN mm)
0	0	3.875	3.875	
0.5	15.5	17.375	21.25	Resilience
1.5	19.25	20.25	41.50	
2.5	21.25	22.15	63.65	Toughness
3.5	23.05	–	–	

Source Author

Unlike the material softening response, the above tabulated data show that the toughness of the given material in the case of the sample-2 is three times the value of its resilience (=area up to the LoP, taken approximately equal to its peak load). The elastic modulus/stiffness of the material can be computed by measuring the slope of the tangent drawn at the origin of the curve. It is given below:

$$= \frac{15 - 0}{0.5 - 0} = 30\text{kN/mm}$$

On the similar lines, the slope of the curve in the post-peak region can be determined; it is given below:

$$= \frac{23 - 20}{3.5 - 1.9} = 1.875\text{kN/mm}$$

The positive slope in the post-peak range is an indication of the material hardening response. Unlike the sample-1, the positive slope indicates that the beam sample-2 is not losing its load carrying capacity after reaching its maximum capacity. The activation of some new mode of the stress transfer mechanism after the previous ones fails to provide the required resistance results in the strain-hardening in the material. This process also results in the higher member toughness; it increases to 63.65 Joules for the sample-2 in comparison with the sample-1 where it was 36.825 Joules. This is the biggest plus point of designing any composite for the strain-hardening phenomenon; the material toughness increases manifold and possesses a higher value of the design stress. In case of strain-softening materials, we have to use a higher value of the factor of safety to cover up loss of the strength in the post-peak region; e.g., in the concrete, a factor of safety of 1.5 is prescribed by most of the guidelines, whereas it is only 1.15 in case of steel (think, why?).

4.3.3 Fatigue

Sometimes, many materials have to perform under repeatedly fluctuating/cycling type of the loading conditions, such as members supporting lifting or moving loads, and/or all those members who are subjected to repeated stress cycles from the vibrating machinery, vehicular movement, wind-induced oscillations, crowd-induced oscillations, etc. In all such scenarios, most of the materials fail prematurely well below their yield strength. The highest stress that any material can withstand for a given number of loading cycles without breaking denotes their fatigue strength. Fatigue is a measure of the weakening of a material caused by the cyclic loading that results in progressive and localized structural damage in the form of movement of dislocations (like in the metals) and initiation of other microcracks (in the composites), especially at the point of the stress concentrations or any other points of the weakness present in them. Once these cracks have been initiated and/or dislocations have started their movement, each loading cycle applied to the materials helps them grow and move further by small amounts. The dislocations and cracks continue to grow until they reach a critical size responsible for a complete fracture of the material component. Because of this, the capacity of the materials to resist the imposed loading goes on reducing as the number of loading cycles is increased. However, some materials like steel sustain a certain stress level even after the lapse

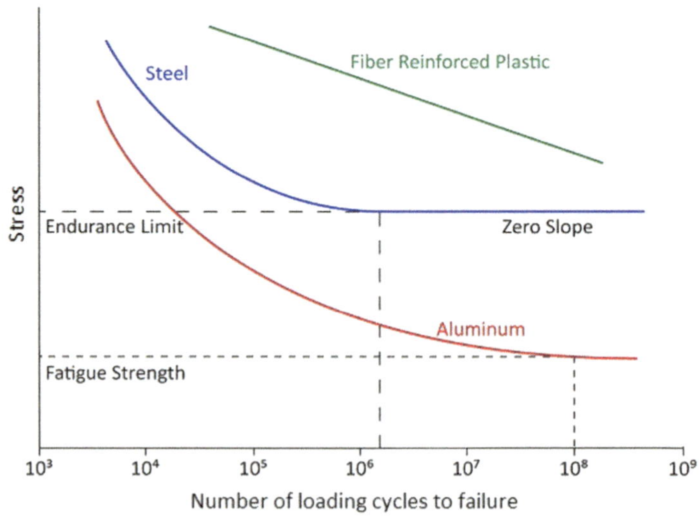

Fig. 4.2 A typical S–N curve for iron and nonferrous metals. *Source* Author

of infinite loading cycles due to the pinning action of interstitial elements present in the material microstructure, such as carbon or nitrogen present in the microstructure of steel, pin the dislocations and restrict them to move further. The stress level below which an infinite number of the loading cycles can be applied to the material without causing any form of the failure is called endurance limit. This limit represents a stress level below which the material does not fail and can be cycled infinitely. If the applied stress level is kept below the endurance limit of the material, the structure is said to have an infinite life. The fatigue behavior of materials is generally expressed by means of a strength-loading cycle (S–N) response curve. One typical such curve for three different types of materials is shown in Fig. 4.2. It shows that some materials, such as aluminum and fiber-reinforced plastics, do not have a well-defined endurance limit as is exhibited by the steel.

Like aluminum, there are many other nonferrous metals such as magnesium and copper that also do not exhibit well-defined endurance limits. In such cases, the fatigue strength is generally defined corresponding to a certain number of imposed load cycles. For instance, the value of the effective endurance limit for these materials is defined as a stress magnitude that causes failure at 1×10^8 loading cycles. The ratio of the endurance limit of a material to its ultimate strength is called the fatigue ratio. Depending on their type, different materials posses fatigue ratio ranging from 0.25 to 0.60. The low value of the fatigue ratio indicates that if any structural member is subjected to cyclic loading, it will need a heavier section to support the load due to the reduction in the value of the safe stress than otherwise.

The S–N curve for most of the materials if plotted on a log–log scale translates into a straight line. A power law equation defines the so obtained linear S–N relationship; it is known as Basquin's equation. It is given in Eq. 4.2, which can be used for the ferrous metals (like the steel) in the loading cycle range of 1×10^3 to

1×10^6. It can also be used for the loading cycle range of 1×10^3 to 5×10^8 for the other commonly used nonferrous metals, such as aluminum.

$$s = A(N)^B \tag{4.2}$$

where the parameter (s) is the stress level that can be applied corresponding to the number of cycles (N) applied to any material/member and the constants A and B can be determined as below; the constant (B) is known as Basquin slope.

Taking 'logs' of Eq. 4.2, log (s) = log (A) + B log (N).

For a small number of stress reversals, N = 1000 (say), stress (s) in the material can be approximated to 90% (say) of its ultimate strength (s_u):

$$\log(0.9s_u) = \log(A) + 3B \tag{4.3}$$

And, for the infinite material life, the stress in any material must be limited to its fatigue strength (=endurance limit x yield strength), which normally happens at $N = 10^6$ in case of the steel. Therefore, Eq. 4.2 modifies to

$$\log(s_e) = \log(A) + 6B \tag{4.4}$$

Subtracting Eq. 4.4 and Eq. 4.3 gives the value of the constant (B) as below:

$B = \frac{\log S_e - \log 0.9 S_u}{3}$. And Eq. 4.4 reduces to the following value of the constant (A):

$$A = \frac{s_e}{10^{6B}}$$

Knowing Basquin'sconstants (A and B), Eq. 4.2 enables the analyst to determine the number of cycles (N) that he needs to cause failure of the material specimen being subjected to a stress level (s). Alternatively, the stress level can be computed that the material can sustain corresponding to any applied number of stress cycles, after which the member will fail due to the fatigue.

The fatigue can be classified as low cycle fatigue and high cycle fatigue. The extents to which materials deform and plastify during the cyclic loading distinguish between the two classes, and the magnitude of the applied stress level during the process is used to decide the transition between the two types. The low cycle fatigue is characterized by repeated plastic deformations in each loading cycle and is associated with the lower numbers of the imposed loading cycles, whereas the elastic deformation defines the start of high cycle fatigue that the material would exhibit.

Example: A steel sample has yield strength of 250 MPa. It is required in the fabrication of some member temporarily required to withstand 50,000 and 2,00,000 number of the stress reversals produced by the given loading conditions. What safe value of the stress should be adopted in the design process?

The endurance limit of steel (=0.6) can be used to determine the stress level below which it would exhibit infinite life. This value of the stress (without any factor of safety) = 150 MPa (=0.6 × 250). For the steel, it corresponds to around 10^6 numbers of cycles.

As the given structural member is being used temporarily, we can adopt the stress level corresponding to 50,000 stress reversals approximately equal to 225 MPa (say, 90% of the yield strength). With this data, Basquin'sconstants (*A* and *B*) can be determined as follows:

$$B = \frac{\log 150 - \log 225}{3} = -0.0586$$

$$A = \frac{150}{10^{-0.3516}} = 337.05$$

Using Basquin's equation (Eq. 4.2), the stress (s) corresponding to the given number of stress cycles (N = 50,000) can be determined:

$$s = 337.05(N)^{-0.0586} = 178.78\,\text{MPa}$$

Similarly, for N = 2,00,000, the stress (s) can be computed as below:

$$s = 337.05(N)^{-0.0586} = 164.84\,\text{MPa}$$

A suitable factor of safety should be applied (usually, taken equal to 1.5 in the working stress method) to get the value of the corresponding design stress.

4.3.4 Crushing

All materials irrespective of their nature, viz. brittle or ductile, crush when these are subjected to excessive compressive loadings. The only difference will be the value of the ultimate strain at their failure. The failure strain will be a small value for the brittle type of materials, whereas it will be relatively very high in the case of the ductile ones; it generally varies between the values of 0.0001 to 0.0002 for most of the brittle materials and can be as high as 0.005 for the ductile materials, e.g., the steel fiber-reinforced concrete. For the natural materials like stones, pebbles, etc., it is not possible to test the specimens the same way as we are testing the concrete cubes/cylinders. Their round shape hinders their testing in the conventional ways. For instance, they will crush prematurely when tried to test under some compressive loading. The stress concentration caused by the applied load at the point of contact, usually very small area due to their round shape, leads to their premature failure, and it generally results their compressive strength in err. An index is therefore used to quantify their crushing strength, known as crushing value.

The aggregate crushing value gives a relative measure of the resistance of an aggregate to crushing under a gradually applied compressive load. The stone aggregates, whether used in the road construction or as aggregates in concrete, primarily transfer the load from one aggregate to the other adjoining aggregates by means of the direct contact they have among each other. The stress transfer is usually through a small area of their contact. The porous or poor aggregates are liable to be crushed to a finer fraction if the load is of a high magnitude applied over them. Therefore, limiting their crushing to a certain value of the finer content is a better approach rather than to determine the true value of their crushing strength. The plus point of the approach will be that it does not require cutting the rocks to certain desirable shape and size, nor it needs any other form of processing to make them fit for testing. Moreover, the finer content so produced in the aggregates during the testing in their raw form has a tendency to fill the open spaces in between them in the test mold and it tends to increase the load carrying capacity of the material to some extent; however, the excessive crushing will have an opposite effect to reduce the strength. So, a limit is put on the extent of the maximum allowable crushing that does not have any adverse effect of the strength.

Instead of a single specimen, therefore a certain quantity of the material is taken for testing purpose under a specified set of the test conditions while determining the crushing value of the stone aggregates, etc. Normally, the aggregates in the surface-dry conditions, passing 12.5-mm IS sieve and retained on 10-mm IS sieve, are taken in the standard cylindrical mold. The aggregates are tamped in the three equal layers in the mold using a round edge rod 25 times each. The dimensions of the mold are given in Table 4.3. A standard plunger is used to apply a compressive load of 40 tonne magnitude gradually over a time period of ten minutes to the aggregates placed in the mold. At the end of the loading, the residual aggregates are sieved again to know the content passing 2.36-mm IS sieve. This weight expressed as percentage of the initial weight taken in the mold is known as aggregate crushing value. The smaller the crushing values of aggregates the better they are. Usually, it is limited to a value of 30% for the aggregates suitable for any concreting work. In the road construction project, it can be as high as 40% for the aggregates to be provided in the bituminous macadam and the subgrade layers.

Another index known as ten percent fine value is sometimes used to quantify the crushing strength of aggregates, especially for the aggregates, such as soft, porous or large-sized ones not coming in the preview of the crushing value test. It attempts to measure the magnitude of load required between 1 and 40 tonne that produce ten percent fines passing 2.36-mm IS sieve. The aggregates in the surface-dry

Table 4.3 Standard test mold dimensions

Parameter (mm)	150-mm diameter mold	75-mm diameter mold
Internal diameter	152.0 ± 0.5	77.0 ± 0.5
Height	$130 - 140$	$70 - 80$
Thickness	16 (min)	8 (min)
Plunger diameter	150.0 ± 0.5	75.0 ± 0.5

Source https://archive.org/details/gov.in.is.2386.4.1963

conditions, passing 12.5 mm sieve and retained on 10 mm sieve, are tamped in the three equal layers in the mold using a round edge rod 25 times each. The aggregates placed in the standard cylindrical mold (115-mm diameter and 180 mm high) are pushed continuously for 10 min at a uniform rate so as to cause a total plunger penetration of 15 mm for the rounded aggregates, 20 mm for the normal crushed aggregates and 24 mm for the honeycombed aggregates. The load is released after reaching the maximum penetration, and the residual material is removed from the cylinder and sieved on a 2.36-mm IS sieve. The fines passing the sieve are weighed, and this weight is expressed as a percentage of the weight of the test sample. Normally, this percentage should fall within the specified range 7.5–12.5, but if it does not, a further test shall be conducted at some other load magnitude adjusted so as to bring the percentage fines within the range of 7.5–12.5%. The mean percentage fines at this load value shall be used in Eq. 4.5 to calculate the load required to produce the ten percent fines:

$$\text{The load required for the ten percent fines} = \frac{X + 14}{Y + 4} \tag{4.5}$$

In Eq. 4.5, the parameter X is the magnitude of the load applied to the sample in tonnes and Y is the mean percentage the fines passing 2.36-mm IS sieve from the two tests at the X tonnes load. The load required to produce 10% fines shall be reported to the nearest whole number if the load applied was 10 tonnes or more, and it will be to the nearest of 0.5 tonnes if the load was of less than 10 tonnes.

In case of a rock query site where the aggregates are yet not available, the crushing strength of the parent rock can be determined by crushing the standard rock specimens to check their suitability for further processing and use as a construction material. The sample from which the test specimens are to be prepared should be taken from a freshly quarried rock and only from the pieces which show no evidence of incipient fracture or planes of any weakness. In such a case, a minimum of three specimens needs to be tested; otherwise, four specimens shall be tested, of which two shall have the planes of weakness, if any, at the right angle to the load axis. The test specimens should be cut and grind in the form of a cylinder having a mean diameter 25 ± 0.5 mm and the height of 25 ± 0.5 mm from the parent rock piece of minimum dimensions 80 × 40 × 40 mm. While preparing the test specimen, it must be checked that the diameter of the cylinder should vary by not more than 0.25 mm and its height by not more than 0.15 mm from that of a standard test specimen. The rock shall not be subjected to any treatment (such as chipping with a hammer) that may cause unintentional damage to the specimen. Flow of cold water should be ensured throughout the grinding, drilling and sawing operations, to avoid any damage arising from overheating of the specimens during their cutting and grinding operations.

After preparing the test specimens, these should be dried for four hours in the oven at a temperature of 100–110 °C and then cooled to the ambient room temperature before applying any loading. The loading is applied axially to the test specimens at a specified rate of 50 kN/min till it crushes. The crushing strength of

Table 4.4 A typical range of the crushing strength for various rock types

Rock type	Strength (MPa)
Basalt	200–350
Dolerite	240–320
Dolomite	50–150
Gneiss	80–330
Granite	90–230
Limestone	50–240
Sandstone	40–200
Shale	35–110
Siltstone	24–120

Source Author

the test specimens is expressed as MPa or kg/cm^2 by dividing the load at the failure and their cross-sectional area. This value of the strength depends upon many factors, such as the extent of weathering to which the rock has been undergone; number of cracks and joints present in the specimen; the nature of the infilling material if any; and the presence of the rock-to-rock contacts across the joints, planarity and continuity of seams and foliations in the specimen. Because of so many factors, there exists a huge variation in the crushing strength being exhibited by the rock specimens. These inherent limitations go in favor of conducting the crushing value test of aggregates prepared from such rock type to see their suitability for construction projects. Table 4.4 lists a range of crushing strength exhibited by common rocks.

Example: A 5 kg sample of coarse aggregates passing 12.5 mm and those retained on 10-mm IS sieve is taken from some construction site. It is filled in the standard mold as per the prescribed procedure; the sample including the mold weighs 6.36 kg. After applying the prescribed value of the compressive load, it is sieved through 2.36 mm sieve; the residue is found to weigh 116.43 g. Compute the aggregate crashing value (ACV), and then, check the suitability of the aggregates for use in some road construction.

The aggregates are suitable for the intended purpose, if its crushing value is found to be less than 30%. It can be calculated by dividing the residue weight (on 2.36 mm) and initial weight of the sample, expressed as percentage.

The weight of the sample = 6.36 − 5.0 = 1.36 kg = 1360 g.
Therefore, the crushing value $= \frac{116.43}{1360} \text{x} 100 = 8.56\%$

The sample crushing value is less than 30%. It indicates that the aggregates are strong enough to bear and transfer the applied stress safely through them to the underlying media. As such, it is acceptable for the intended purpose.

Example: A 5 kg sample of coarse rounded aggregates passing 12.5 mm and those retained on 10-mm IS sieve is taken from some construction site. It is filled in the standard mold as per the prescribed procedure; the sample including the mold weighs 6.96 kg. The load is continuously applied over the sample; a load of 15 tonne produces the required penetration of 15 mm. The sample in the mold is again sieved through 2.36 mm sieve; the residue is found to weigh 135.63 g. Check the suitability of the aggregates for use in any road construction.

Another way to determine the suitability is to find the ten percent fines of the sample. From the given data, it can be easily calculated.

The weight of the sample = 6.96 − 5.0 = 1.96 kg = 1960 g.

The fines passing the 2.36 mm sieve as percentage of the initial sample weight $= \frac{135.63}{1960} \text{x} 100 = 6.92\%$. This value must fall in the prescribed range of 7.5–12%; otherwise, the test must be re-conducted on a fresh sample.

The load required for the ten percent fines $= \frac{X+14}{Y+4}$, where X is the load applied to the sample (=15 tonne) and Y is the mean percentage of the fines passing 2.36 mm sieve (=6.92%). This value is computed as 2.65 tonne; it should be reported as 3 tonne as the applied load is more than 10 tonne. The value should be compared with the approved specifications to see the suitability of their use. If this value is less than the approved value, the sample should be accepted.

4.3.5 Impact

Unlike the previous indexes, this value gives a relative measure of the resistance of aggregate to some sudden shock or impact. While determining the impact value, the test specimen is subjected to a suddenly applied load instead of the gradually imposed load as was applied in the crushing value determination. The aggregate passing 12.5-mm IS sieve and those retained on 10-mm IS sieve are filled in the standard cylindrical mold (of 102-mm diameter and 50 mm high). The aggregates comprising the test sample should be dried in an oven for a period of four hours at a temperature of 100–110 °C and then cooled before conducting any further test. A weight of 13.5–14 kg is made to have a free fall from a height of 380 ± 5 mm, thus imparting the impact energy to the aggregates placed in the mold. The test specimen is subjected to a total of fifteen such blows, each being delivered at an interval of not less than one second. Depending upon their composition and strength, the aggregates crush under the impact of the applied load; the residual material is taken from the mold and sieved through 2.36-mm IS sieve. The fraction passing the sieve should be weighed to an accuracy of 0.1 g. It should be checked that if the weight passing and that retained on the 2.36-mm IS sieve is less than the weight of the sample taken in the mold or not, and if it is found yes, the whole procedure shall be repeated afresh on another set of the aggregates till this condition is satisfied. The fraction of the aggregates passing 2.36 mm sieve, expressed as a percentage to the weight of initially oven-dried sample taken in the mold, gives their impact value. The mean of at least two samples shall be used to indicate the aggregate impact value; it should be expressed to the nearest whole number. The aggregates possessing impact value up to 10 are exceptionally strong, and those

Table 4.5 Maximum allowable aggregate impact value of different types of paving materials

Types of pavement material/layer	Aggregate impact value, %
Subbase course and water-bound macadam (WBM) layer	50
WBM base course with bitumen surfacing	40
Built-up spray grout, base course	40
Wet mix macadam (WMM) base course and WBM surface course	30
Dense bituminous macadam binder courses	30
Bituminous surface dressing, carpet and bituminous concrete surface	30
Cement concrete surface course	30

Source https://archive.org/details/gov.in.is.2386.4.1963

having values up to 30 are permitted for use in any type structural concreting work. Table 4.5 gives their maximum permitted values for different types of road construction activities. If the impact value of the test specimens is found more than the permitted values, the lot of that aggregates is not suitable for use in the wearing course.

Example: A 5 kg sample of coarse aggregates passing 12.5 mm and those retained on 10-mm IS sieve is taken from some construction site. It is filled in the standard mold as per the prescribed procedure; the sample including the mold weighs 7.16 kg. A load of 14 kg is given free fall as per the prescribed procedure, and the sample is sieved through 2.36 mm sieve; the residue is found to weigh 125.15 g. Compute the aggregate impact value (AIV), and then, check the suitability of the aggregates for use in the concrete production.

The aggregates are suitable for the intended purpose, if its impact value is found to be less than the tabulated value (see Table 4.5). It can be calculated by dividing the residue weight (on 2.36 mm) and initial weight of the sample, expressed as a percentage.

The weight of the sample = 7.16 − 5.0 = 2.16 kg = 2160 g.

Therefore, the aggregate impact value= $\frac{125.15}{2160} \times 100 = 5.79\%$.

The value is significantly less than the requirement of 30%; hence, the sample can be used for the intended purpose.

4.3.6 *Abrasion*

Abrasion is the wear and tearing of the top exposed surface of the materials because of the rubbing action of another moving substance over it. For applications of materials such as in the floorings, pavements and road surface, we need materials that could resist this type of the deterioration. The surface roughness of materials mobilizes frictional forces between the two moving substances against each other that led to the loss of material from their top contact surfaces. Depending upon their bulk and surface properties, they try to resist the tearing away of the particles/ constituent blocks near their top contact surfaces. The stronger the bond in the material microstructure the higher will be their resistance to the abrasion. Alternatively, it can be decreased by reducing the generation of frictional force; e.g., the vitrified flooring tiles provide seamless surface finish. The abrasion mainly results in the localized loss of material from the contact surfaces and the loosening between the aggregate and the paste in case of the cement-based composites, such as concrete. A number of factors, including material compressive strength, the properties of aggregates, water/cementitious ratio, the presence of other supplementary materials like silica fume, fly ash or fibers in the mix, control the abrasive resistance of concrete. Any approach by which we are able to estimate the loss of the material from the contact face of a specimen is fit to assess the abrasion. Based on this concept, a number of the methods have been standardized and are in use to calculate the relative resistance offered by materials to abrasion. A material undergoing a relatively smaller loss of its topping is more resistant to the abrasive action than the others. The test results are usually reported in the form of abrasion coefficients, namely average abrasion depth and weight loss. Two of the most common methods in use are described below:

The first method attempts to measure the loss of the material from a test specimen caused by the abrasive action of air-driven silica sand used in the testing operations. ASTM C418-05 prescribes a method known as sand blast abrasion testing that aims to determine the abrasion resistance of the cement-based composites, such as concrete in the form of an abrasion coefficient (A_c). The coefficient can be obtained by dividing the volume of the material lost from the specimen due to the abrasive action and its abraded surface area. The reduction in the height of the test specimen—circular disks of dimensions 150 × 64 mm—and its other dimensions during the testing can be measured by means of Vernier caliper, and it can be converted to the abrasion coefficient.

Another method uses test specimens in the form of a revolving disk to determine the abrasive resistance of materials. It is prescribed by ASTM C779-05. In this method, frictional forces are induced on the surface of test specimens through rubbing and grinding action of rotating steel disks in conjunction with abrasive grit. The disks are free floating, transversely driven along a circular path at 12 rpm, while at the same time, they individually turned on their own axis at 280 rpm. The cups attached to the top of the shaft of each disk are loaded with the lead shots to produce a uniform total compressive load of 22 N on each abrading disk face. The silicon carbide abrasive (No. 60) is fed to the disks at a rate of 4–6 g/min. The test is carried out for a duration of 30 min. The reduction in the height of test specimens is measured by means of Vernier caliper, and this value is used to calculate the weight loss of the specimen and the average abrasion depth.

IS 1237 and IS 15658 also prescribed an abrasion testing procedure to determine the abrasive resistance of concrete specimens, such as concrete paver. Here, the square-shaped test disk of size 71.0 ± 0.5 mm is cut from the concrete block to be tested for abrasion. It must be ensured during the cutting operations that the contact face and the opposite face of the disk should be parallel and flat. The test specimens are dried in oven at a temperature of 105 + 5 °C, and their density is determined nearest to 0.1 g. The weight of the specimen should be noted to the nearest of 0.1 g both prior to the abrasion test and after every four cycles of the testing. The grinding path of the testing disk of the abrasion testing machine should be evenly strewn with 20 g of the standard abrasive powder. The disk is fixed in the holding device such that the testing surface faces the grinding disk. The specimen is centrally loaded with a force of 294 + 3 N. The grinding disk is then run at a speed of 30 rpm. The disk shall be stopped after one cycle of 22 revolutions. The disk and contact face of the specimen are cleaned of abrasive powder and debris. The specimen is then turned 90° in the clockwise direction, and 20 g of abrasive powder is evenly strewn on the testing track before starting the next cycle. The test cycle shall be repeated 16 times, the specimen being turned 90° in the clockwise direction and spreading of 20 g of abrasive powder on the testing track after each cycle. The abrasive wear of the specimen after 16 cycles of testing is computed as the mean loss in the test specimen volume (taken in mm^3) by dividing the loss in the mass (in g) of the specimens after 16 cycles and its density (in g/mm^3).

Example: A concrete circular disk of size 150-mm diameter and 64 mm high is tested as per the provisions of ASTM C418-05. The height of the sample is measured as 147.23 mm after the test. Compute the value of abrasion coefficient.

The coefficient can be determined by computing the volume of the material removed from the sample during the test and its abraded surface area.

The change in the height of the sample = 150 − 147.23 = 2.77 mm.

The volume of the material removed during the process = $\frac{3.14\times150^2\times2.77}{4} = 48{,}925.125\,\text{mm}^3$

Abrading area = $\frac{3.14\times150^2}{4} = 17{,}662.5\,\text{mm}^2$

This gives the value of the coefficient as 2.77.

4.3.7 *Permeability*

Porosity and permeability are intrinsic properties of every material. The microstructure of a material greatly controls its porosity and the consequential permeability. The porosity refers to the amount of empty pore space that exists within the material, while the permeability is a measure of their connectedness and defines how easily a fluid flows through a porous material. If the material has a high permeability, then the pore spaces therein are connected to one another, allowing fluids to flow from one to another. On the other hand, a low permeability indicates that the pore spaces are isolated and/or small enough that the fluids remain trapped within them. For example, in the loosely packed gravels because of their large size all of the pores in them are well connected among each other, thereby allowing the water to flow through it very easily, which becomes very difficult in case of a clayey soil stratum. The extremely small-sized clay particles remain bonded to each other because of the presence of electrostatic charge on their surfaces, which also keeps the water molecules in the form of moisture trapped therein. As a result, such type of the soils and other similar materials exhibit low permeability; they do not allow the water to seep through them. Overall, it is the relative size of the pore size and the configuration of seeping liquid molecules that define the permeability characteristics of any material. The permeability is the capacity of the materials to transfer liquids through a fully saturated pore network under a hydrostatic pressure gradient. The water first tries to fill the pores and saturate the materials. Once that has happened, the pressure gradient across opposite faces of the materials forces the pore water to move through the interconnected pore structure. The permeability plays a key role in the selection of suitable building materials for different construction activities. For instance, we cannot choose a material of high permeability for the storage of water or in the floors of building, especially its roof. And, some material having a high permeability will be the best if we have to use it in the open parking areas where the purpose is to keep the area free of water accumulation during the rainy days. Table 4.6 lists the permeability coefficients for the commonly found soils; it varies from the almost negligible value (e.g., metals) to very high values (e.g., well-sorted pack of gravels).

The permeability of materials can be easily computed if we maintain a pressure difference across the opposite faces of a test specimen and allow the water to flow through its thickness in a certain time interval. IS 3085 [5] prescribes a procedure to determine the permeability of the cement-based composites. The test specimens are prepared in the form of cylindrical shape, having a height and the diameter of 150 mm each. However, in the case of specimens containing aggregates whose nominal size does not exceed 20 mm, the dimensions of the specimen can be reduced to 100 mm and where the nominal aggregate size exceeds 40 mm, the dimensions should not be less than about four times the nominal size of the aggregate.

The specimen should be cast in the split molds of the required size and be cured for 28 days before testing them under a standard test pressure of 10 kg/cm^2. This may, however, be reduced up to 5 kg/cm^2 in the case of the materials (e.g., highly porous in nature) where it is possible to obtain the steady-state conditions in a reasonably shorter time, and the pressure head may be increased up to 15 kg/cm^2 in the case of the materials having a low permeability. It should be ensured that no water should be allowed to leak from the sides of the specimens when these are duly sealed in the test mold. The open annular space around the specimen should be filled using the cotton or hemp cord soaked in a suitable molten sealing compound, completely level with the top of the specimen; otherwise, the leakage through the seal if any can give rise to entirely misleading results.

Once the setup is ready, the next part is to apply the hydraulic pressure gradient across the specimen faces and just observe the drop in the pressure. The test should run continuously for 100 h. The flow should be permitted to attain the steady state before the coefficient of permeability is calculated from Eq. 4.6. Examination of the

Table 4.6 Typical value material permeability of different soils

K (cm/s)	10^{2}	10^{1}	1	10^{-1}	10^{-2}	10^{-3}	10^{-4}	10^{-5}	10^{-6}	10^{-7}	10^{-8}	10^{-9}	10^{-10}
Range of relative permeability	Pervious				Semi-pervious				Impervious				
Unconsolidated sand, gravel and other building materials	Well-sorted gravel		Well-sorted sand or mix of sand and gravel, pervious concrete		Very fine sand, silt, loess, loam, bricks, normal concrete					High-strength concrete, clayey soils, polymers and metals (toward the 10^{-10})			
Unconsolidated clay and organic soil	-				Peat		Layered clay			Non-weathered clay			
Rocks	Highly fractured rocks				Oil reservoir rocks			Sandstone		Limestone and dolomite		Granite	

Source Author

inflow and outflow rate data or suitable graphs of the same may be used to determine the establishment of the steady state.

$$k = \frac{Q}{At\frac{H}{L}} \tag{4.6}$$

In the above equation, k is the coefficient of permeability in cm/sec; Q is the quantity of water in milliliters percolating over the entire period of the test after the steady state has been reached; A is the area of the specimen face in cm^2; t is the time in seconds over which Q is measured; and ***H/L*** denotes the ratio of the pressure head to the specimen thickness.

Example: A standard disk of 150-mm diameter and 150-mm height is used to determine the permeability of the concrete specimen. It is tested under a pressure of 10 kg/cm^2. Under this pressure, 1250 ml of water passes through the specimen in 105 h. Determine the coefficient of permeability of the test specimen.

Equation 4.6 can be used to estimate the coefficient of permeability of the concrete specimen. The area of the test specimen is A = 176.62 cm^2.

The pressure head used during the testing operations is $H = \frac{100 \times 10^3}{1000} = 100\,\text{m}$; therefore, the term H/L in the expression 4 is computed as 666.67 (=100/0.15).

The term $Q/t = \frac{1250}{105 \times 60 \times 60} = 3.31 \times 10^{-3}$.

The coefficient of permeability (k) of the concrete specimen $= \frac{Q}{At\frac{H}{L}} = \frac{3.31 \times 10^{-3}}{176.62 \times 666.67} = 2.8 \times 10^{-8}$ cm/s.

4.3.8 Time-Dependent Properties, Such as Shrinkage and Creep

Almost all materials undergo changes over a period of time; these can be in their dimensions, the load carrying capacity or their texture. It all depends upon the composition and their microstructure; some materials do it rapidly, while for many others, it is not possible to observe these minute changes that take place in them during the lifespan. For instance, the cement-based composites exhibit shrinkage just after (or during) their casting operations, while it is almost insignificant for the stones and the metals. This type of the time-dependent material deformations sometime gets magnified under the influence of externally applied load. Most interestingly, these types of stress-magnified deformations occur over and above those caused by the shrinkage.

Shrinkage is a shortening of a material with time. Generally, the materials having a relatively more pore space and the presence of high specific area molecules in their microstructure exhibit this type of the response over a certain period of time once they have been produced/cast. For instance, clay in the moist state shrinks as the moisture entrapped in the material evaporates. As the water molecules leave the

Table 4.7 Autogenous shrinkage for different concrete grades

Concrete grade	Autogenous shrinkage strain, $\times 10^{-6}$
M30	35
M35	45
M45	65
M50	75
M60	95

Source https://archive.org/details/gov.in.is.1343.2012

clay microstructure, maybe due to evaporation or otherwise, the secondary bonding among the different molecules tends to bring them closer, thereby exhibiting shrinkage. On the other hand, the sand will not shrink being the coarser material than the clay particles. Similarly, the cement-based composites like concrete undergo significant shrinkage owing to their unique microstructure. It contains pore spaces and also holds water molecules therein. The metals and polymers on the other hand posses very fine grain structure having pore spaces that are much finer than the size of the water molecules. Because of this characteristic, these two materials remain impervious to most of the liquids and did not shrink as the concrete does. Changes of pore water content due to drying or wetting processes cause significant volume changes of concrete specimens. The corresponding value of the strain ranges between 0.0002 and 0.0005 in case of the shrinkage and around 0.00005 in case of the swelling. All those factors that influence the evaporation from any surface also control the shrinkage; the relative humidity is one of the major factors that control it. In addition, the total shrinkage in concrete depends upon the constituents of concrete and the member sizes. For a given humidity and temperature conditions, the total shrinkage of concrete is mostly influenced by the total amount of water present in the concrete at the time of mixing and, to a lesser extent, by the cement content.

The strain in concrete caused by shrinkage is composed of two components, the autogenous shrinkage strain and the drying shrinkage strain. The shrinkage of the concrete during its hardening stage is classified as autogenous shrinkage strain. The major part of this strain component develops during the early days after its casting and it go on reducing with time. The value at any time (***t***, in days) can be computed by multiplying the tabulated shrinkage (Table 4.7) and the decay function $[= 1 - e^{-0.2\sqrt{t}}]$. The strain is observed to be a linear function of the strength of the concrete used to cast the members; a typical set of the shrinkage values for different concrete grades are given in Table 4.7 for a ready reference.

On the other hand, the drying shrinkage occurs slowly, as it is a function of migration of the water through the pores as they escape to the atmosphere in the hardened concrete. The size of the member plays a major role in controlling the drying shrinkage. A parameter called as notional size (h_o) is used to quantify the effect of member dimensions. It can be calculated (in mm) as $2A_c/u$, where A_c is the concrete cross-sectional area (in mm^2) and u is the perimeter (in mm) of that part of the cross

Table 4.8 Drying shrinkage values* ($\times 10^{-6}$) for concrete produced with Portland cement

Concrete grade	Relative humidity, RH 50%	Relative humidity, RH 80%
M25	535	300
M50	420	240
M75	330	190

*For other values, linear interpolation is permitted
Source https://archive.org/details/gov.in.is.1343.2012

section which is exposed to drying. A set of typical values of drying shrinkage is given in Table 4.8. These values must be multiplied by the factor (k_h) tabulated below depending upon the value of the notional size of the concrete member:

h_o (mm)	k_h
100	1.0
200	0.85
300	0.75
≥ 500	0.70

Source https://archive.org/details/gov.in.is.1343.2012

Similar to the autogenous strain, the drying shrinkage in concrete also decays with time. It will be the maximum at the end of the curing period (concrete age t_o in days) and then decays as the time passes. The value at any time (t, in days) can be computed by multiplying the value taken from Table 4.8 with the decay function given in Eq. 4.7.

$$\beta(t, t_o) = \frac{(t - t_o)}{(t - t_o) + 0.04\sqrt{h_o^3}} \tag{4.7}$$

The results of drying shrinkage in the section can be computed by multiplying the value of the corresponding strain with the decay coefficient β (t, t_o) and the constant k_h. The total strain on account of the shrinkage process in any concrete member will be the sum total of the autogenous and the drying shrinkage strains being induced therein at any given set of the relative humidity (RH) and the member geometric dimensions.

Creep is another important material property that badly affects the structural performance of members cast using concrete or other similar materials if its effects are not properly considered in the design of members. Unlike the shrinkage strain, the presence of an external load on the member is a necessary condition to begin the creeping process. It is basically a forced evacuation of the entrapped moisture from the pores of the concrete, which leads to further shortening of the member over and above the value caused by the shrinkage. The paste in the concrete has a porosity of about 0.4–0.55, and it poses an enormous internal surface area, around 500 m^2/cm^3 which traps moisture in the pores. Some part of this moisture escapes by means of the

Table 4.9 Creep coefficient φ_0 (t = 70 year, t_o) of an ordinary structural concrete after 70 years of loading

Age at loading t_o (days)	RH: 50%			RH: 80%		
	Notional size (mm)			Notional size (mm)		
	50	150	600	50	150	600
1	5.8	4.8	3.9	3.8	3.4	3.0
7	4.1	3.3	2.7	2.7	2.4	2.1
28	3.1	2.6	2.1	2	1.8	1.6
90	2.5	2.1	1.7	1.6	1.5	1.3
365	1.9	1.6	1.3	1.2	1.1	1.0

Source https://archive.org/details/gov.in.is.1343.2012

evaporation process, while the other parts try to move out under the action of external compressive stresses. During this process, the concrete undergoes shortening in their dimensions. This type of the straining in the material is known as creep. It can be determined by finding the change in the dimension that a specimen under the effect of sustained loading will undergo. It is mandatory to ensure that the loading must be maintained at a constant level; it actually reduces as the member shortens so we have to adjust the imposed loading to the preset load levels during the testing operations.

A number of expressions and procedures [1, 6] are available to estimate the strain that the material creep will produce. It is generally expressed by means of the creep coefficient, defined as a ratio of the creep strain and the initial stain in the member caused by the imposed load. The creep strain can be computed by multiplying the notional creep coefficient (φ_0) to whom the creep coefficient reaches asymptotically with time (usually, it happens in about 70 years) and another coefficient β (t, t_o) that describes the development of the creep with time. The value of the notional creep coefficient can be taken from Table 4.9. The value of the coefficient β (t, t_o) is given in Eq. 4.8.

$$\beta(t,t_o) = \left(\frac{(t-t_o)}{(t-t_o)+b}\right)^{0.3} \tag{4.8}$$

The parameter (t) denotes the age of the specimen/member at the time considered in the analysis, whereas t_o is the concrete age at the time of the loading. The value of the parameter (b) in the equation depends upon the relative humidity (e.g., RH of 80% taken as 0.8 in the expression), the notional member size (h_o in mm) and the concrete grade. It can be calculated from the following expressions:

$$b = \begin{cases} 1.5\left[1+(1.2RH)^{18}\right]h_o + 250 \le 1500 & \text{for} \quad \mathrm{f_{ck}} \le 45\,MPa \\ 1.5\left[1+(1.2RH)^{18}\right]h_o + 250\alpha \le 1500\alpha & \text{for} \quad \mathrm{f_{ck}} > 45\,MPa \end{cases}$$

where $\alpha = \left(\frac{45}{f_{ck}+8}\right)^{0.5}$.

Unlike the cement-based composite materials, the metals normally undergo creep at elevated temperature. The mechanism behind the creep in the metals is also different from that causes the concrete to creep. The metals exhibit creep deformations mainly because of the grain boundary sliding that took place in them under sustained loading conditions. The movement becomes fast with the rising temperature and the metals depending upon their grain size and chemical composition starts exhibiting plastic deformations. Since creep deformation occurs by grain boundary sliding, the more the grain boundary area, the easier and the higher the creep deformation will be. Therefore, the creep response of the metals can be improved by developing a larger grain size in them during their production or by alloying them with some other elements that could pin the movement of dislocations in the grains.

Example: A 75-mm-thick concrete slab having plan dimensions of 1.5 m × 2.5 m is cast using M30 concrete. It is water cured for 7 days and then left for air curing. Determine the change in the dimensions that will occur after 14 days from the casting date. The RH during these days can be taken as 50%.

The change in the length or its width can be determined easily if we are able to find the strain caused in the slab by the shrinkage process. It composed of two major components, namely autogenous shrinkage strain and drying shrinkage strain. Their values can be easily computed from the data provided in the problem.

The autogenous strain $\varepsilon_a = 35 \times 10^{-6}$ for M30 concrete (ref. Table 4.7). It will decay to some new value during the 14 days from the day of the casting $= \varepsilon_a\left(1 - e^{-0.2\sqrt{14}}\right) = 18.49 \times 10^{-6}$.

The value of the drying shrinkage strain for M30 concrete can be taken as 512×10^{-6} (ref. Table 4.8). Its decay with time will be controlled by the geometric dimensions of the slab and the time lapsed from the cessation of the curing period.

The notional depth (h_o) of the slab along 2.5 m side $= \frac{2(1500\times75)}{(2\times75+1500)} = 136.36\,\text{mm}$. It will give the value of the coefficient (k_h) = 0.945 for this side of the slab. Similarly, the coefficient for the other dimension (i.e., 1.5 m) of the slab is computed as 0.938.

The decay coefficient for the drying shrinkage strain $\beta(14, 7) = \frac{(14-7)}{(14-7)+0.04\sqrt{136.36^3}} = 0.099$.

The value of the drying shrinkage ε_d (14, 7) = $0.099 \times 0.945 \times 512 \times 10^{-6} = 47.91 \times 10^{-6}$.

The total shrinkage in the slab along its 2.5 m length = $\varepsilon_a + \varepsilon_d = 18.49 \times 10^{-6} + 47.91 \times 10^{-6} = 6.64 \times 10^{-5}$.

Therefore, the expected shortening in the slab = $2500 \times 6.64 \times 10^{-5} = 0.166$ mm. In case the slab is restrained to shrink freely, it leads to the development of the tensile stress in the slab section of magnitude = εE = 1.81 MPa. If the slab concrete section has tensile strength of a value more than the induced tensile stress, it will not crack; otherwise, shrinkage cracks will form in the section. Similarly, the total shrinkage that the slab section will experience along its 1.5 m side can be calculated as 0.0955 mm.

Example: A 150 mm concrete cube is cast using M30 concrete. It is water cured for 7 days and then left for air curing. At the age of 28 days, it is placed under a compressive force of 50 kN for many days. Determine the change in the dimensions of the cube that will occur at 90 days from the casting date. The RH during these days can be taken as 50%.

As the concrete cube is placed under the compressive force, it will experience straining on account of both the shrinkage andcreep. Their respective values can be computed by using the empirical relations presented above in the section.

The autogenous strain $\varepsilon_a = 35 \times 10^{-6}$ for M30 concrete (ref. Table 4.7). It will decay to some new value during the 50 days from the day of the casting = $\varepsilon_a\left(1 - e^{-0.2\sqrt{90}}\right) = 29.79 \times 10^{-6}$.

The value of the drying shrinkage strain for M30 concrete can be taken as 512×10^{-6} (ref. Table 4.8). Its decay with time will be controlled by the geometric dimensions of the slab and the time lapsed from the cessation of the curing period.

The notional depth (h_o) of the specimen$= \frac{2(150\times150)}{(4\times150)} = 75\,\text{mm}$. It gives the coefficient (k_h) = 1.

The decay coefficient for the drying shrinkage strain $\beta(90,7) = \frac{(90-7)}{(90-7)+0.04\sqrt{75^3}} = 0.762$.

The value of the drying shrinkage ε_d (50,7) = $0.762 \times 1 \times 512 \times 10^{-6} = 389.94 \times 10^{-6}$.

The total shrinkage in cube = $\varepsilon_a + \varepsilon_d = 29.79 \times 10^{-6} + 389.94 \times 10^{-6} = 4.197 \times 10^{-4}$.

The creep strain can be computed from the elastic strain caused by the imposed loading by multiplying it with the corresponding value of the creep coefficient at the time t = 90 days. The age of the concrete at the time of its loading t_o = 28 days.

Initial elastic strain $= \frac{\sigma/A}{E}\frac{50000}{150\times150\times5000\sqrt{30}} = 8.11 \times 10^{-5}$.

The creep coefficient = 2.4 for the notional size of 75 mm (ref. Table 4.9).

The corresponding decay in the strain = $\beta(90,28) = \left(\frac{(90-28)}{(90-28)+362.51}\right)^{0.3} = 0.561$.

Therefore, the creep strain at this age $= 8.11 \times 10^{-5} \times 0.561 \times 2.4 = 1.09 \times 10^{-4}$.

The total strain in the cube can be estimated by finding the sum total of the strains caused by the shrinkage and the creep = $4.197 \times 10^{-4} + 1.09 \times 10^{-4} = 5.29 \times 10^{-4}$. It will translate to the specimen shortening of 0.08 mm ($=150 \times 5.29 \times 10^{-4}$).

4.3.9 Durability Considerations

A durable material is one that performs satisfactorily in the working environment during its anticipated exposure conditions during the entire service life. Like the mechanical properties, the chemical composition and microstructure of the material also control their durability. The pore size and their distribution in materials affect their permeability characteristics; a high value leads to an easy ingress of water and other gasses like oxygen, chloride, carbon dioxide, sulfate and other potentially deleterious substances into the material microstructure and the consequential deterioration, especially the cement-based composites suffer the most on this account. The chemical composition on the other hand affects the life of the metals; they are prone to corrosion if not properly protected against using some means. Because of this, the metals start losing their geometrical dimensions, thereby affecting the load carrying capacity. The wooden structural members are very prone to the termite attack which also led to loss of their geometric dimensions, thus influencing their load carrying capacity. The response of the polymeric materials, like the metals, is highly dependent on the temperature. They start losing their mechanical properties as well as exhibiting relatively higher creep deformations at the elevated temperatures.

It is relatively easy to manage the metals, wood and polymeric materials than the concrete or the concrete-like materials. A protective coating or alloying will serve the purpose to minimize the chances of material to lose because of either rusting or termite in the case of the metals and wood/timber, respectively. But in case of the concrete, the only choice is to densely pack the microstructure by filling the pore spaces. The impermeability in the concrete specimen is governed by the constituents and workmanship to obtain adequate compaction and efficient curing. The size and shape of the concrete member and the cover are provided to the embedded reinforcing steel; the type and quality of constituent materials, the cement content and water–cement ratio used in the production of the concrete should be carefully selected to make concrete perform better throughout its service life. Any lapse generally leads to the corrosion of the embedded reinforcing steel in concrete members. It is all natural for the steel to convert to its more stable form if exposed to moist air in open, or the moist air, otherwise, finds its way to the steel rebars embedded in concrete members. Once that happens, it destroys the protective layer formed on the surface of embedded reinforcing bars in the concrete member by lowering the prevailing pH from 13 to less than 9. It is only the concrete microstructure that protects the embedded steel from corroding by providing imperviousness. An attempt should therefore be made by appropriately selecting the mix proportions to make the concrete as dense as possible, which helps to reduce its permeability and finally helps to enhance the durability of concrete members.

The presence of chlorides in concrete also increases the risk of corrosion of embedded steel. The higher the chloride content (or even if, subsequently, the concrete with chloride content, even in small quantities, is exposed to the warm, moist conditions), the greater the risk of the corrosion of the embedded reinforcing steel will be there. Many constituents may contain chlorides and also, the concrete may be contaminated with the chlorides from the external environment. In order to minimize the chances of deterioration of concrete from chlorides or other similar salts, their levels either coming into the concrete from the constituent materials, such as cement, aggregates, water and admixtures, or by diffusion from the environment should be limited to 0.4 kg/m^3 of concrete.

Some aggregates containing particular varieties of silica are susceptible to attack by alkalis, such as Na_2O and K_2O originating from cement or other sources in the concrete. The resultant reactions being expansive in nature usually leads to internal cracking in the concrete, which finally translate to spalling or disintegration generally observed in the concrete members. An attempt, therefore, should be made to use the aggregates that do not contain alkali reactive constituents in their material composition.

Concrete otherwise also should not undergo an average weight loss of more than one percent after having been subjected to 50 freeze–thaw cycles while totally immersed in a 3% sodium chloride solution. It can be determined in the controlled laboratory conditions. The equipment consists of three cabinets to complete the freezing–thawing operations. The freezing cabinet must have a capacity to reach and maintain an air temperature of -15 ± 2 °C within 1 h of the introduction of the test specimens, while the thawing cabinet should be suitable to maintain a controlled air temperature of 23 ± 3 °C. The moist cabinet should be capable to maintain a controlled air temperature of 23 ± 2 °C and a relative humidity of 90%. One freeze–thaw cycle should be completed every 24 h. It consists of 16 h of the freezing, followed by 8 h of the thawing. If for any reason a thaw period cannot be started at the specified time, the specimens should be kept in a frozen condition until the conditions become suitable for resumption of the test.

After bringing the specimens to the room temperature, these should be placed in individual containers with the bottom surface of the specimens resting on the glass or plastic spacers to ensure exposure of at least 95% of the bottom surface of the test specimen to the saline solution prepared from three percent sodium chloride solution. The level of the solution shall be maintained at least 2 mm above the top surface of the specimens, but the excess volume of the solution shall be avoided in order to ensure rapid freezing of the specimens. The specimens should be kept in the solution for 24 h getting them fully saturated and then shall be subjected to continuous freeze–thaw cycles. After every 10, 25 and 50 cycles, the specimens should be washed with three percent sodium chloride solution to remove any loose particles. These particles and spalled material if any, collected at the bottom of the containers, shall be washed, strained through a filter and then dried. This residue dried mass of the material constitutes the weight loss and is expressed as a percent of the initial dry weight of the specimens. The residue shall be cumulatively weighed after 10, 25 and 50 cycles. The weight loss shall be calculated to the nearest 0.01%. The test should be continued until 50 freeze–thaw cycles have been completed, unless the test specimens have either disintegrated or there is a loss of more than 1.0% of the original dry weight in the specimens.

4.4 Different Methods to Monitor the Quality of Materials

The quality of materials being produced either on-site or at some distant mechanized plants continuously changes. Numerous factors influence this parameter on which the performance of any component or the structure as a whole depends. A variation more than the permitted controls or those listed in the approved acceptance criterion leads to underperformance or development of inadequate strength in the affected members. Some methods have been devised in this regard and are frequently used at the sites to ensure that any material being used in or going into some construction activity conforms to the approved norms.

The first method known 'within-batch precision' approach compares the variations between any two batches, and if it is found to be within the permitted limits, the batch is allowed to be used at the site; else, the problem is diagnosed and some necessary corrective actions are taken to rectify the defects at the earliest possible. In this approach, the strength of a sample (maybe the compressive strength of concrete or the tensile strength of reinforcing steel bars) consisting of a minimum of 10 specimens (such as cubes) per lot is determined; the number of samples that can be taken from the batch can be two, three or four. The approximate deviation (s_1) that exists in the test results from two consecutive batches can be calculated by dividing the range (R) of the data by a factor (d) as given below for a different number of samples:

No. of samples	Factor, d
2	1.128
3	1.693
4	2.059

Source Author

Knowing the sample deviation (s_1) and the average value (X) of the test results, the within-batch coefficient of variation (V_1) can be determined from the following expression:

$$s_1 = \frac{R}{d}$$

$$V_1 = \frac{s_1}{X} 100$$

This way of assessing the suitability of the material being brought to the site for use is an easy and quick approach. On-site decisions can be taken on the basis of the coefficient, V_1. The batch is acceptable, if the coefficient of variation (V_1) is found up to four or maximum five.

Example: Consider the following set of test results obtained from a batch of concrete. The test results in the form of concrete compressive strength are tabulated below and are used to find the coefficient of variation within the batch.

Sample-1 results (MPa)	Sample-2 results (MPa)	Average strength (X) (MPa)	Range (R) of the test results (MPa)
67.40	71.20	69.30	3.8
70.50	67.50	69.00	3.0
56.40	58.30	57.35	1.9
55.70	55.50	55.60	0.2
60.30	57.00	58.65	3.3
56.90	56.50	56.70	0.4
55.30	56.00	55.65	0.7
53.50	53.20	53.35	0.3
46.50	50.80	48.65	4.3
58.00	60.80	59.40	2.8
	Average	58.37	2.07

Source Author

Deviation, $s_1 = 2.07/1.128 = 1.835$.

Coefficient of variation, $V_1 = 1.835 \times 100/58.37 = 3.14\%$.

As the observed variation within the batch from two different samples is around 3%, it can be considered acceptable and the batching and the concrete production operations do not need any corrective action. If the value of the coefficient (V_1) is found larger than 6%, the concrete should not be allowed to pour and the batching needs immediate corrections to rectify the defects. The coefficient of variation (V_1) up to 5% is acceptable for most of the concrete work.

Another method also works on a similar principle, but it needs continuous data from the site to monitor the quality of the material being used in different construction activities. This method is known as CUSUM approach as it uses the cumulative sum of the deviations that occurs in the test sample from its average value to see the variations in the material properties. A graph is plotted between the CUSUM and the sample number to see the trend in the test results. If the batch is of a uniform quality, then the CUSUM will be of zero slopes; otherwise, it will fluctuate up and down depending upon the quality of the material being to the site for use. If a steep negative slope is noticed anywhere in the process, it is indicative of a poor quality material finding its way to the site. In such a case, immediate alert should be issued to the supplier and/or batch plant persons to get the things right. This process is more practical than the former approach and easy to use to get a live feedback about the material quality. The following examples will illustrate the working of this approach:

Example: The cubes from some construction site are tested regularly, and their compressive strength (in MPa) data are listed here: 24.89, 30.22, 25.78, 26.22, 23.11, 22.22, 21.33, 27.11, 23.56, 24.0, 20.0, 21.33, 26.22, 30.22, 26.67, 28.89, 30.22, 31.11,22.67,23.11,21.78, 24.44, 25.33, 26.67, 28.89, 31.11, 17.33, 27.56, 29.78, 19.11, 25.78, 21.33, 24.89, 25.33, 30.22, 31.56, 27.56, 28.44, 24.89, 23.11, 25.78, 21.33, 17.33, 21.78, 25.33, 21.33, 20.0, 21.33 and 26.22. It pertains to various footings being poured at the site. We have to monitor the consistency of the test results and whether it is acceptable if the grade of the concrete is M20.

In the CUSUM approach, the test data will be processed to compute the deviation in the reported values from the average of the sample. The process is tabulated below:

Cube no.	Reported cube compressive strength (x) (MPa)	Deviation from the sample mean ($x - x_m$) (MPa)	CUSUM value (MPa)
1	24.89	−0.253	−0.253
2	30.22	5.077	4.824
3	25.78	0.637	5.461
4	26.22	1.077	6.538
5	23.11	−2.033	4.505
6	22.22	−2.923	1.582
7	21.33	−3.813	−2.231
8	27.11	1.967	−0.264
9	23.56	−1.583	−1.847
10	24	−1.143	−2.99
11	20	−5.143	−8.133
12	21.33	−3.813	−11.946
13	26.22	1.077	−10.869
14	30.22	5.077	−5.792
15	26.67	1.527	−4.265
16	28.89	3.747	−0.518
17	30.22	5.077	4.559
18	31.11	5.967	10.526
19	22.67	−2.473	8.053
20	23.11	−2.033	6.02
21	21.78	−3.363	2.657
22	24.44	−0.703	1.954
23	25.33	0.187	2.141
24	26.67	1.527	3.668
25	28.89	3.747	7.415
26	31.11	5.967	13.382
27	17.33	−7.813	5.569
28	27.56	2.417	7.986
29	29.78	4.637	12.623
30	19.11	−6.033	6.59
31	25.78	0.637	7.227
32	21.33	−3.813	3.414
33	24.89	−0.253	3.161
34	25.33	0.187	3.348

(continued)

(continued)

Cube no.	Reported cube compressive strength (x) (MPa)	Deviation from the sample mean ($x - x_m$) (MPa)	CUSUM value (MPa)
35	30.22	5.077	8.425
36	31.56	6.417	14.842
37	27.56	2.417	17.259
38	28.44	3.297	20.556
39	24.89	−0.253	20.303
40	23.11	−2.033	18.27
41	25.78	0.637	18.907
42	21.33	−3.813	15.094
43	17.33	−7.813	7.281
44	21.78	−3.363	3.918
45	25.33	0.187	4.105
46	21.33	−3.813	0.292
47	20	−5.143	−4.851
48	21.33	−3.813	−8.664
49	26.22	1.077	−7.587
Sum total	1224.42		

Source Author

The average of the sample $x_m = \frac{1224.42}{49} = 24.99$ MPa.

The deviation of the reported strength value from the sample average is given in the table and it is cumulatively summed up. The plot of this value (CUSUM) is plotted below (see the graph below) to observe the variation in the consistency of reported results.

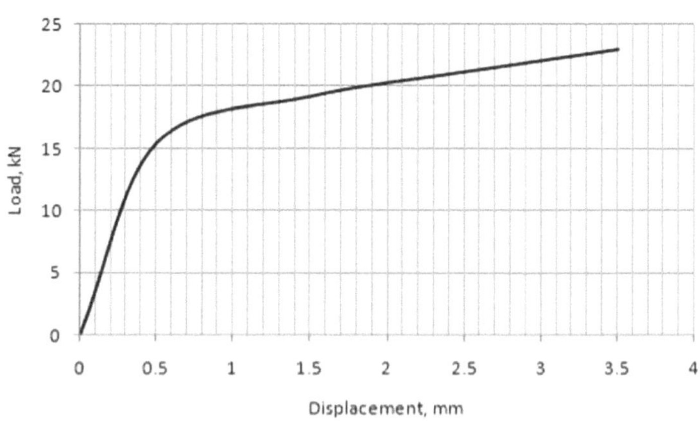

The variations in the cube compressive strength are reflected in the graph. It indicates that the concrete quality starts degrading corresponding to the 5th cube taken from some concrete pour and it goes on degrading further till the 15th cube. When such continuous degradations are observed in the plotted data, the site officials are required to alert the concrete supplier and/or the batch plant operator to take the corrective measures. Once these are taken, the improvement in the quality will start reflecting in the graph; i.e., the curve starts depicting a positive slope as is indicated in 12th cube value. A concrete of consistent quality will always depict a curve with almost uniform plus–minus trend. The controls should be so set to get this type of the variation in the test data. It is impossible to get a straight horizontal line curve for any material (think, why?). A similar problem is again noticeable when the concrete sample in 40th cube is coming for testing and analysis. If some required corrective action is not taken, the concrete quality will go on degrading as is reflected in the graph and the problem may be out of control when the CUSUM becomes negative as is occurring in 46th cube number.

It is interesting to note that the average value of sample compressive strength is around 25 MPa and it appears satisfactory for the M15 concrete grade. But the huge variations in the strength from cube to cube are very high, and it is sufficient to trigger failure at the point of weakness created by the use of poor quality material at any specific point in the member. This type of monitoring process helps to identify any such points in the member being poured and take suitable remedial actions to remove the defects in material (concrete) production process. This process is very easy to implement at the site; a spreadsheet will come very handy to see the day-to-day variation in the strength properties of any material being used at the sites. The graph will go on depicting the variations when some new test data are entered in the spreadsheet. Any negative slope needs immediate attention and should not be ignored.

Another way of evaluating the material quality is by means of characteristic strength. It will ensure that any concrete batch will have only five percent probability of failure in the entire sample. It will need calculation of the sample mean (x_{m}) and the standard deviation (σ) of the reported data.

For the test data given in the above table, the sample mean is calculated as 24.99 MPa and the standard deviation can be calculated from Eq. 2.2.

$$\sigma = \sqrt{\frac{(x - x_m)^2}{n-1}} = \sqrt{\frac{653.71}{48}} = 3.69\,\mathrm{MPa}$$

The characteristic strength (f_{ck}) of the sample = $24.99 - 1.64 \times 3.69 =$ 18.93 MPa. This value is coming less than what is considered in the design of concerned member (i.e., 20 MPa). This indicates non-acceptance of the material for the said construction. This value must be more than the design characteristic strength of the material. The shortfall can be improved by modifying the production process producing the material with a smaller standard deviation in the test results. For instance, the value of the standard deviation in this case must be limited to 3.0 MPa only to get the desired concrete compressive strength (check, why?).

4.5 Nondestructive Testing Procedures

Nondestructive testing provides an alternative to the conventional testing procedures where the quality of the material used in construction or products is required to be assessed without damaging them. These also help to evaluate their strength properties. The range of properties that can be assessed using nondestructive tests is quite large and includes density, elastic modulus and strength as well as member surface hardness, surface absorption, reinforcement size, location and its distance from the concrete surface. Nondestructive testing can be conducted on both old and new structures. The purpose however is entirely different in each case; in case of the new construction, it is for a quality control or the resolution of doubts about the quality of materials being used in the construction, while it is the assessment of structural integrity or adequacy in case of existing buildings/structures. Typical situations where nondestructive testing may be useful are listed below:

- Quality control of precast units or in-situ construction activities
- Removing uncertainties about the acceptability of the material supplied to the sites owing to some apparent non-compliance with the specifications
- Confirming or negating doubt concerning the workmanship involved in batching, mixing, placing, compacting or curing operations
- Monitoring of strength development in relation to formwork removal, cessation of curing, prestressing, load application or similar purpose
- Location and determination of the extent of internal cracks, voids, honeycombing or other similar defects within a concrete structure
- Determining the concrete uniformity, possibly before the actual conduct of core cutting or load testing or other some more expensive testing operations
- Determining the position, quantity or condition of embedded reinforcement in concrete
- Increasing the confidence level when a smaller number of destructive tests are to be (or have been) conducted
- Determining the extent of concrete variability in order to help in the selection of sample locations representative of the quality to be assessed
- Confirming or locating suspected deterioration of concrete resulting from factors such as overloading, fatigue, external or internal chemical attack, fire, explosion and environmental effects
- Assessing the potential durability of the concrete
- Monitoring long-term changes in concrete properties of existing structures
- Providing information for any proposed change of use of a structure for insurance or for change of ownership/use.

There are a number of well-established nondestructive procedures that can be used to evaluate the above-mentioned construction-related needs. Visual inspection is the very first step in the process that helps to plan and identify which of the various available methods could be the most useful for any further investigation of the problem. Some of the methods are listed below that are commonly employed in the investigations:

- Schmidt/rebound hammer test, used to evaluate the surface hardness of concrete
- Ultrasonic pulse velocity testing mainly used to measure the homogeneity of concrete by means of the sound waves. It also helps to detect voids or other internal cracking in the concrete and give a fair idea about the concrete compressive strength
- Carbonation depth measurement test finds application to determine whether moisture has reached the depth of the reinforcing bars
- Half-cell electrical potential method is used to detect the corrosion potential of embedded reinforcing bars in concrete
- Permeability test is used to measure the flow of water through the concrete, thus giving some idea about the porosity of the material
- Covermeter testing, used to measure the distance of embedded steel reinforcing bars from the surface of a concrete member and their diameter.

4.5.1 Rebound Hammer Test

The rebound hammer, also called as Schmidt's hammers, works on the principle that the rebound of an elastic mass depends on the hardness of the surface against which it impinges. Basically, it is an indicator of the surface hardness of a material. However, various empirical relationships have been developed and are available that can be used to estimate the compressive strength of the material from the measured rebound number. Usually, a high rebound number represents concrete with a higher compressive strength than the concrete with a low rebound number. This happens because of a significant influence of aggregates used in the production of concrete on its elastic properties. Therefore, it should be used only if some correlation is available or can be developed between the rebound number and the concrete made with the same type of coarse aggregates as that are being tested. Generally, a rebound number of more than 30 indicate a good concrete surface and blow 20, a porous surface. If this value is coming near zero, the chances are that the test surface is delaminated or cracked internally, maybe because of possible rebar rusting in case of reinforced concrete. It is necessary that the rebound hammer is checked against the testing anvil having Brinel hardness of 5 kN/mm before starting the actual testing and the value so obtained must confirm to the data sheet provided by the supplier/manufacturer of the rebound hammer [3].

Because of its small size and low weight, the Schmidt hammer is suitable for both the field and the laboratory investigations. It consists of an outer body, the plunger, the hammer mass and the main spring. Other features include a latching mechanism that locks the hammer mass to the plunger rod and a sliding rider to measure the rebound of the hammer mass after striking a surface being tested. The rebound distance is measured on an arbitrary scale marked from 10 to 100, which is recorded as a 'rebound number' corresponding to the position of the rider on the scale. Three variants of the hammer are available with different impact energy ranges, including Type L (0.75 Nm), Type N (2.25 Nm) and Type M (29.45 Nm). Depending upon the anticipated concrete strength, the hammer with an appropriate impact energy level should be selected to get a reliable estimate of the concrete compressive strength. Additionally, a proper care should be taken to minimize the possible errors arising on account of the following factors:

Smoothness and rigidity of the specimen: The test surface has to be smooth and firm in order to get a true value of the rebound number. There should not be any loose patch or material on the test surface. If the thickness of the member is small, it may lead to a reduction in the rebound number as the member may deflect or deform under the impact and dissipate certain amount of the imparted energy in the process. In such cases, the member has to be rigidly held or should be backed up by placing a heavy mass behind it to get the most reliable results. It should not be carried out on the concrete surface during its early ages or when the concrete strength is less than 7 MPa since the concrete surface could be damaged by impact of the hammer in all such cases.

Surface and internal moisture conditions of concrete: The rebound number gets affected by the presence of moisture in the concrete. These are found to be on the lower side for the well-cured air-dried specimens in comparison with the results obtained from an identical set of the specimens tested in the wet conditions or tested in their saturated surface-dried conditions after removing the excess water from the surface. Therefore, whenever the actual moisture condition of the field concrete or any test specimen is unknown, the surface should be pre-saturated for several hours before testing. A correlation curve for tests performed on saturated surface-dried specimens should then be used to estimate the compressive strength.

Type of coarse aggregate: Aggregates control the elastic properties of concrete. A concrete produced from soft aggregates or having a smaller elastic modulus will show the smaller value of the rebound number than the concrete otherwise. Therefore, an appropriate correlation should be arranged or be developed between the rebound number and the concrete made with the same type of coarse aggregate as that are being tested.

Orientation of the hammer: The hammer can be used in the horizontal, vertically up or down positions as well as at any intermediate angle, provided it is held perpendicular to the surface being tested. During the testing, the position of the mass relative to the vertical affects the rebound number. The hammer will report the value smaller if any member (e.g., some floor) is tested from its top face than on its soffit or some other inclined orientations due to the action of gravity on the mass in the hammer. Therefore, an appropriate correlation curve should be used to convert the rebound number to the concrete compressive strength.

Carbonation of the concrete surface: The rebound number is found to be affected by the depth to which the concrete surface is carbonated. The carbonated concrete surface produces around 50% lower rebound number in comparison with the concrete surface otherwise. A porous concrete specimen or the concrete having extensive internal microcracking will, therefore, exhibit lower rebound number than a concrete mass which remains uncarbonated, maybe because of dense microstructure or the presence of some protective layer/coating, etc.

4.5.2 Pulse Velocity Testing

The solids exhibit unique elastic properties depending upon their chemical composition and the microstructure. Some are very dense, and there are many which possess relatively loose microstructure. There are many solids that have inherent internal flaws in the form of voids, cavities, or microcracking, e.g., concrete. So when sound waves are made to pass through the solids, there exists a wide range of velocities with which the sound wave travels for each and every case. In case of a very fine microstructure (for instance, in the metals), these travel very fast, whereas it is relatively slow in the case of the concrete. It will be very slow if the concrete is extensively cracked or have a very loose microstructure. These variations in the sound wave velocities enable the analyst to detect the material homogeneity and the presence of internal cracks therein.

Pulse velocity measurements therefore are easily used for quality control purposes. Unlike the mechanical tests on control samples such as cubes, this type of the testing has the advantage that they relate directly to the concrete actually used in the structure rather than to rely on the samples, which may not be always truly representative of the in-situ concrete. Generally, the wave frequency in the range of 20–150 kHz serves the intended purpose of quality control and detection of other defects in materials. Because of this frequency range, the pulse velocity testing is also known as ultrasonic pulse velocity testing (UPV test) [2]. The high-range frequencies are usually employed for measurements over the short path lengths, such as in study of mortar thickness, grouts and thin slab/plates, whereas the low frequency should be used in the investigations when some long path lengths exist, e.g., in case of the beams, slabs, columns, etc., having large one dimension in comparison with their other sides.

UPV equipment consists of an electrical pulse generator, a pair of transducers, an amplifier and an electronic timing device for measuring the time interval between the transducer placed at the point of pulse generation and another transducer held at the receiving side. When the pulse is induced into a material from a transducer, it may undergo multiple reflections at the boundaries of the different phases present in the materials. Usually, a complex system of stress waves in the form of longitudinal, transverse and surface waves is generated in the material. The longitudinal waves being the fastest are normally used in the testing and the receiving transducer that is calibrated to detect their arrival only and transform the data into the velocity form. In case of concrete, the velocity of the sound waves gives a fair idea about the material quality; Table 4.10 enlists a range of the velocities usually observed in the concrete depending upon the in-situ density and its corresponding material quality classification to judge the conditions of the concrete test specimen.

Table 4.10 Velocity criterion for assessing the quality of concrete

No.	Pulse velocity (km/s)	Quality of the concrete
1	>4.5	Excellent
2	3.5–4.5	Good
3	3.0–3.5	Medium
4	<3.0	Doubtful

Source https://archive.org/details/gov.in.is.13311.1.1992

The velocity measurements through a material can be made by placing a pair of transducers on the opposite faces of the concrete member. This placement, known as direct transmission or cross-probing, ensures a transfer of the maximum possible energy across the member without any loss, and it should be preferred over any other possible arrangements. For instance, while evaluating the concrete quality used in the casting of the beam or column section, the transducers can be placed on their opposite faces. However, there are many cases where it cannot be done, e.g., in case of the slabs/floors; the transducers is such cases has to be placed on its soffit only as the top face will be usually hidden under the flooring. This type of a transducer placement is called as 'indirect transmission' or 'surface probing'; it should be used only when one face of the concrete is accessible or when the purpose of the investigations is to determine the quality of the surface concrete relative to the overall quality or estimate the depth of possible cracking from that face of the member. It is worth to note that the quality of the signal recorded by the transducer in indirect transmission is generally less than five percent of what will be recorded during the direct transmission mode. It is always, therefore, advisable to make the measurements using the direct transmission as far as possible.

A series of measurements are made along a line on the surface of the member with the transmitter fixed at some position while the receiver transducer moving along the line and recording the time of arrival of the signal. The transmission times are plotted as points on a graph showing their relation to the distance separating the transducers. The slope of the straight line drawn through these points is measured and recorded as the mean pulse velocity along that line on the concrete surface. Wherever any break or change in the slope is noticed in the graph, it is indicative of a presence of some surface crack or layer of inferior quality in that area. It is sometimes possible to place the transducers on opposite faces of the member but not directly opposite to each other, for instance, transducers placed on adjacent faces of the member, maybe the sides and the bottom of some beam or two adjacent faces of a column. Such an arrangement of the transducers is regarded as semi-direct transmission. It gives results in between the former two placement positions. It is however always desirable to assess the concrete quality by using all the three transducer placement positions appropriately. A typical transducer (T) and receiver (R) positions are shown in Fig. 4.3 for all the three possible positions of sensors. An attempt should be made to get the data from all possible positions; e.g., both the direct and semi-direct transmissions should be used in the assessment of concrete quality in the beams; one set of pulse velocity should be obtained by placing the transducers on the vertically opposite faces of the beams, and another

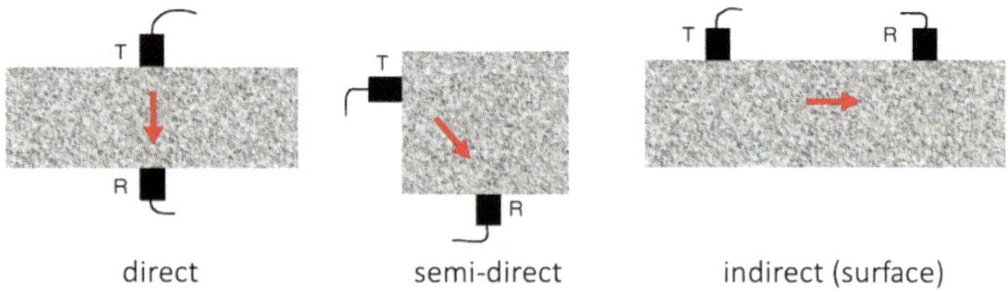

Fig. 4.3 Different transducer positions that can be used while conducting UPV testing. *Source* Author

set of data can be obtained by placing one transducer on the bottom face of the beam while the other on the vertical face of the beam web. If the quality of the concrete is consistent and it is more or less homogeneous in composition, the results from both placement positions will come on the similar lines; otherwise, there may be a significant variation in the pulse velocities.

Ideally, pulse velocity should be used to see the variations in the concrete quality along and across the member length. However, this velocity data can also be used to assess other defects, such as cracks, and material properties of the in-situ concrete by developing correlation between the measured pulse velocity and the desired concrete property, maybe its compressive strength and elastic modulus. Such correlations can be readily established directly for precast units and can also be found for in-situ concrete at the sites. One such relationship between these elastic constants and the velocity of an ultrasonic pulse traveling in an isotropic infinite elastic medium is given in Eq. 4.9.

$$E = \left[\frac{(1+\mu)(1-2\mu)}{(1-\mu)}\right]\rho v^2 \tag{4.9}$$

In the above equation, E is the dynamic elastic modulus (in MPa), μ is the dynamic Poisson's ratio, ρ is the density (in kg/m^3) and v is the pulse velocity (in km/s).

Equation 4.9 can be used to estimate the elastic modulus (E) of the in-situ concrete from the known values of its Poisson's ratio and the density along with the pulse velocity data obtained from that concrete specimen. Alternatively, the value of Poisson's ratio (μ) of the concrete can be obtained from the other three values of the parameters in the expression. It is always advisable to determine the concrete density from the test results of the cores taken from the same region of the specimen. The procedure can be used for a wide range of shapes or sizes as the pulse velocity through the member is not significantly affected by its dimensions, except when one or more of the lateral dimensions is small relative to the wavelength of the pulse used in the testing. A typical propagation of sound wave through the concrete member is depicted in Fig. 4.4. It is the fastest along the path shown at the

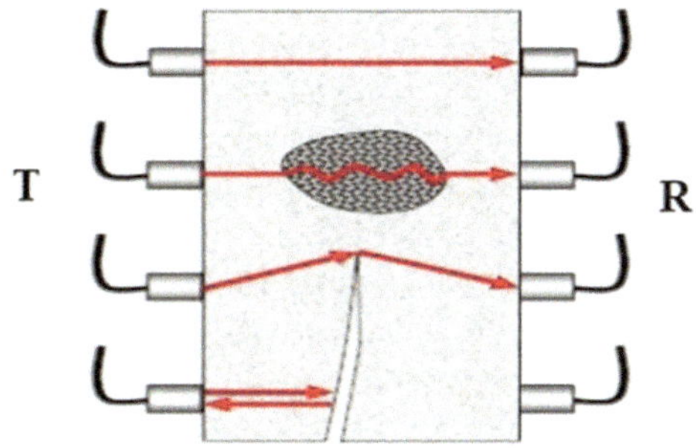

Fig. 4.4 Sound wave propagation through a typical concrete member having some area of looseness and cracking on one of its faces. *Source* Author

top of the figure, while it will take maximum time for the path at its bottom where the sound wave encounters the internal crack. The difference in the time taken along the bottom path and some other path at a distance from the bottom face helps to determine the depth of the crack because after some distance (equal to the crack depth), the pulse velocity will come nearly same for all other positions of the measurements. Similarly, this process can be used to identify the presence of some loose pocket in the concrete and also its size/dimensions. If such types of pockets are observed in large numbers in the member under investigations, it indicates a poor quality of the concreting work.

It should be kept in mind that sound pulse travels at almost double the velocity through the steel than the concrete. It is therefore expected that the pulse velocity in the concrete area measured in the vicinity of embedded reinforcing bars will be higher than in the region away from the reinforcing bars. The apparent increase in pulse velocity depends on the proximity of the measurements to the reinforcing bar, the diameter and number of rebars and their orientation with respect to the wave propagation path. Therefore, a suitable correction may be required to get the true value of the pulse velocity or the measurements should be made in such a way that the steel bars do not lie in or close to the direct path between the transducers.

Further, certain corrections have to be made in the velocity measured using the transducers if the temperature at the time of measurements is not in the prescribed range of 10 and 30 °C; the temperature in this ranges has been found to cause an insignificant effect on the velocity. The percent correction that has to be made in the measured pulse velocity to get its true value is given in Table 4.11; alternatively, the

Table 4.11 Date sheet for the temperature corrections

Temperature (°C)	Correlation made on the measured pulse velocity in	
	Air-dried concrete (%)	Water-saturated concrete (%)
60	+5.0	+4.0
40	+2.0	+1.7
20	0	0
0	–0.5	–1.0
–4	–1.5	–7.5

Source https://archive.org/details/gov.in.is.13311.1.1992

measurements should be done in an ambient room temperature conditions and it should be avoided to the extent possible to test the specimens in freezing temperatures where the measured pulse velocity can go as high as 7.5% in comparison with the room temperature conditions; else, we have to reduce the velocity by the same amount to get the true value of the pulse velocity.

The pulse velocity also changes with the variations in the moisture content of the specimens; the moist concrete will pass the pulse faster than the dry specimen. It can be two percent higher in the saturated concrete than that of a similar specimen of the dry concrete. An attempt should be made to test the concrete having a smooth test surface only, in its surface-dry conditions. To ensure that the ultrasonic pulses generated at the transmitting transducer pass into the concrete to the maximum possible extent and are then detected by the receiving transducer, it is essential that there be adequate acoustical coupling between the concrete and the face of each transducer. In the absence of a smooth surface, couplets such as petroleum jelly and grease may be applied over the test surface to get good acoustical transmission of the pulse. An appropriate transducer should be selected to take the measurements depending upon the travel path and the dimensions of the member. Usually, a frequency of more than 50 kHz serves the purpose to evaluate 100 mm thick specimens over a path length of 1500 m. If smaller thicknesses (less than 100 mm) are to be evaluated, the higher frequencies around 150 kHz should be preferred; it can be used to test a surface as small as 25 mm in thickness.

Example: The UPV testing measures the pulse velocity through a concrete beam as 3.85 km/sec. The core taken from the beam shows a density of 2385 kg/m^3 along with a compressive strength 37 MPa. What should be the estimated value of the concrete elastic modulus (E)?

Equation 4.7 can be used to estimate the value of the concrete elastic modulus. The dynamic Poisson's ratio for the concrete can be assumed as 0.2; otherwise, it can be calculated from the core test data. Knowing all the parameters required in Eq. 4.7, the elastic modulus (in MPa) can be determined as follows:

$$E = \left[\frac{(1+\mu)(1-2\mu)}{(1-\mu)}\right]\rho v^2 = \left[\frac{(1.2)(0.6)}{(0.8)}\right]2385(3.85)^2 = 31,816\text{MPa}$$

Otherwise, if the value of the concrete density and its elastic modulus is obtained from the core testing, the value of the dynamic Poisson's ratio can be computed from Eq. 4.7. These three material properties in addition to the concrete strength constitute an important input data that is needed in various structural analysis programs/computing tools. It also helps engineers take various on-site decisions regarding the possibility of using the building/components for some other purpose than for what they are designed.

4.5.3 Carbonation Depth Measurement

Concrete contains numerous cracks ranging from a few isolated to a network of well connected one. Their thickness varies from few nanometers to millimeters, depending upon the loading conditions and other environmental factors. This makes it possible for the atmospheric gasses and the moisture to move easily in and out of the concrete. This destroys the natural pH (of 12–14) existing the concrete and leads to its reduction to a value less than 9 depending upon the concentration of the gas (mostly, CO_2)/moisture. The area in the concrete where the gasses find an easy access gets carbonated, while the other nearby areas may contain calcium hydroxide. This enables detection of the areas in the members that gets effected and the other areas that remains unaffected. It can be easily done by spraying a freshly exposed surface of the concrete with one percent phenolphthalein solution. The concrete portion containing a lot of the calcium hydroxide becomes pink, while the carbonated portion remains uncolored. The solution can be made by dissolving 1 gm of the phenolphthalein in 90 cc of ethanol. The solution is then made up to 100 cc by adding distilled water.

While determining the carbonation in the concrete, the cores taken from the member are sprayed with phenolphthalein solution. The depth of the core (from the outer face of the surface) that remains uncolored is measured to the nearest millimeter at 5 or 10 positions and the average taken. The test can also be performed in a drilled hole or any exposed surface of the concrete [7]. It is important to remove the dust from the hole or surface using an airbrush and again the depth of the uncolored layer measured at 5 or 10 positions and the average taken. A stable reading may be obtained after about 5 to 10 min after spraying.

The test also finds application to assess the potential of possible rebar corrosion which occurs once pH reduces to less than 9 or so. The carbonated area in the concrete member is most prone to corrosion, so this way of testing helps to identify such areas which are prone to rusting.

Problems

Q1. A standard concrete cube of 150 mm size was tested in compression on a strain-controlled machine. The test data are given below:

Displacement (mm)	Load (kN)
0	0
0.05	13.1
0.5	17.2
1.5	12.2
2.5	9.3
3.5	7.3

Calculate the compression strength of the concrete sample; its modulus of elasticity; design value of the compression strength; toughness and resilience.

Prepare a scientific article on this experiment as per the guidelines shared in the self-study content related to the unit-3. The theory part must cover the related fundamentals, machine description and way of getting such test data.

Q2. Three different materials were tested in the laboratory. The load–displacement responses of these materials are given below:

Displacement (mm)	Load, kN for different samples		
	A	*B*	*C*
0	0	0	0
0.05	13.1	12.6	12.3
0.5	17.2	17.3	24.2
1.5	12.2	19.5	11.4
2.5	9.3	21.4	6.5
3.5	7.3	25.2	4.1

Decide the type of the material from this response data (i.e., strain hardened, strain softened, elastic, plastic, elasto-plastic, etc.). Use spreadsheet to plot the response curves. Calculate the compression strength of each, their toughness, resilience and modulus values (initial, tangent and secant). Based upon these results, comment on the suitability of these materials for different applications. Which should be the best for a routine RC construction?

Q3. Write a short note on crushing value and impact value of aggregates. What are their acceptable values for permitting use of aggregates in concrete construction?

Q4. Describe 10% fine method to determine the suitability of aggregates for construction work. What are the other methods that can be used for this purpose?

Q5. A sample of aggregates weighing 10 kg was sieved; 8.25 kg passed 12.5 mm sieve, and out of this, 6.35 kg was found to be retained on 10 mm sieve. It was placed in a standard mold, 40 tonne load was applied for 10 min, and 2.75 kg of the material was found to retain 2.36 mm sieve. Check the suitability of the material for some road construction work. If not found suitable, suggest how can we use this lot of the material for the said project.

Q6. What do you mean by impervious materials? If some material is, say, very porous in texture, what type of problems it can pose? Explain with the help of suitable illustration.

Q7. Determine the permeability of a material if water is made to flow through the sample in an area of 10 cm diameter under a water head of 5 cm. Assume that all other faces of the samples are duly sealed except this area.

Q8. Differentiate shrinkage and creep exhibited by different materials.

Q9. Why do the man-made materials, such as concrete, exhibit relatively higher shrinkage and creep than the natural materials, such as stone?

Q10. Discuss the importance of shrinkage and creep phenomenon in the design process of different structural elements.

Q11. How can you reduce the effects of creep and shrinkage being exhibited by materials? Enlist different influential factors that control these two material properties.

Q12. Estimate the total shortening of 3-m-long concrete member (made up of M25 grade) when it is cast during the month of April and, then, in August at some location in Ludhiana. The x-sectional size of the member is 350 × 350 mm. The member is expected to carry an axial load of 600 kN on the 40th day of casting. It will be water cured for the initial 5 days. Based upon your results, suggest which month will be more appropriate for the member casting.

References and Bibliography

References

1. ACI Committee 209 (1972) Prediction of creep, shrinkage and temperature effects in concrete structures. In: ACI-SP27, Designing for effects of creep, shrinkage and temperature, Detroit
2. IS 13311–1 (1996) Method of non-destructive testing of concrete, Part 1: Ultrasonic pulse velocity
3. IS 13311–2 (1996) Method of non-destructive testing of concrete-methods of test, Part 2: Rebound hammer
4. IS 2386–4 (2002) Indian standard methods of test for aggregates for concrete: mechanical properties
5. IS 3085 (1965) Method of Test for Permeability of Cement Mortar and Concrete
6. RILEM (1988) Committee TC 69. In: Bažant ZP (ed) Mathematical modeling of creep and shrinkage of concrete. Wiley, Chichester and New York, pp 57–215
7. RILEM Committee TC56 (1988) Measurement of hardened concrete carbonation depth -CPC 18. Mater Struct 21(126):453–455

Bibliography

8. Alexander M, Mindess S (2005) Aggregates in concrete. Taylor & Francis: United Kingdom
9. Atlas of Stress-strain Curves (2002) ASM International, United States
10. Blitz J, Simpson G (1996) Ultrasonic methods of non-destructive testing. Springer, Germany
11. Boullion TL, Duran BS, Bethea RM, Boullion TL (1995) Statistical methods for engineers and scientists. Taylor & Francis, Hong Kong
12. Materials C (2018) Their nature and behaviour, 4th edn. CRC Press, United States
13. Dislocations in Solids (2004) Elsevier Science, Netherlands
14. Durability of Concrete Structures: Investigation, Repair, Protection (1991) CRC Press, United Kingdom
15. Hoff WD, Hall C (2011) Water transport in brick, stone and concrete. CRC Press, United Kingdom
16. Hull JB, John V (2015) Non-destructive testing. Macmillan, United Kingdom
17. Kobayashi T (2004) Strength and toughness of materials. Springer, Germany

18. Krishnadas Nair CG, Prasad J (2008) Non-destructive test and evaluation of materials. Tata McGraw-Hill, India
19. Lewis DM (1951) Magnetic and Electrical Methods of Non-destructive Testing: Report Prepared for the Magnetic and Electrical Methods, Sub-committee of the British Iron and Steel Research Association. Allen & Unwin, United Kingdom
20. Mehta PK, Monteiro PJM (2013) Concrete: microstructure, properties, and materials. McGraw-Hill Education, United Kingdom
21. Nawy EG (2001) Fundamentals of high-performance concrete. Wiley, United Kingdom
22. Non-Destructive Evaluation of Reinforced Concrete Structures: Deterioration Processes and Standard Test Methods (2010 Elsevier Science, India
23. Non-destructive Testing and Evaluation of Civil Engineering Structures (2017) Elsevier Science, Netherlands
24. Performance Criteria for Concrete Durability (1995) Taylor & Francis, United Kingdom
25. Prakash R (2009) Non-destructive testing techniques. New Age Science, United Kingdom
26. Raj B, Jayakumar T, Thavasimuthu M (2002) Practical non-destructive testing. Woodhead, United Kingdom
27. Richardson MG (2003) Fundamentals of durable reinforced concrete. Taylor & Francis, United Kingdom
28. Shah SP (2006) Measuring, monitoring and modeling concrete properties: an international symposium dedicated to professor Surendra P. Shah, Northwestern University, USA. Springer, Netherlands
29. Singh H (2017) Steel fiber reinforced concrete: behavior, modelling and design (Springer Transactions in Civil and Environmental Engineering), Singapore
30. Wadsworth HM (1998) Handbook of statistical methods for engineers and scientists. McGraw-Hill, United Kingdom
31. Yang L, Huang Y, Zhou Y (2013) Micro- and macromechanical properties of materials. CRC Press, United States

Chapter 5
Documenting Test Results

Testing and evaluation of materials is an important part of any quality control process. In order to get the best out of the project, this aspect should be considered as an investment which pays large dividends throughout the project life. A variety of materials go into the construction of any project. Ensuring that a right material finds its way to the site, and its subsequent use during the construction is the outcome of an effective testing strategy. It also aims to ensure that whatever material goes into the construction schedule must meet the prescribed norms to perform satisfactorily its intended functions. An effective testing and evaluation strategy helps to cut the unnecessary expenditure that otherwise would incur on clearing operations at site of rejected/unacceptable materials and on the demolition and recasting operations in case any unacceptable material is used in the construction activities.

Documentation plays an important role to see compliance of the mandatory regulations, both the statuary and the non-statuary. It includes all files and documents that contain information on the testing strategy, day-to-day monitoring and the progress records, achieved results, and evaluation metrics. This data serve to measure the testing effort to achieve the desired project quality at the site along with a control on the test coverage across various horizontals and verticals of the project and also track the future project requirements. As the project moves along, owner; developers and promoters; architects and engineers can come back to the documents any time and see if the actual performance and the progress correspond to the planned activities or not; and if required, necessary course corrections can be done at the earliest possible to get the desired level of control. The direct outcomes of this process ensure a satisfactory performance of the project, its broad acceptability among various stakeholders and the confidence which finally decides the value of the project and helps the promoters to monetize the inputs. This chapter presents methods and procedure to record the test results related to various materials, finding their use in different structural applications.

H. Singh, *Structural Materials*, Materials Horizons: From Nature to Nanomaterials,
https://doi.org/10.1007/978-981-16-3211-2_5

5.1 Test Data

Structural materials are the backbone of any project. The entire skeletal system of the project grows and develops as per the conceptual and working drawings by use of these materials in the execution. It is therefore utmost important that they should be produced, transported and placed at the right place in their right composition. Any deviation or lapse in this chain of construction activities generally leads to underperformance and other unexpected material responses in the members. In order to ensure a satisfactory response while performing their intended functions in the structural members, such as their design capacity, toughness and durability, a variety of testing methods have been established and are in practice. The batches of materials who qualify the mandatory check points find their use in various construction-related applications. In this process, a well-defined, standardized procedure is followed for each and every test parameter, which allows the results of like tests performed by different laboratories to be compared. If these are performed exactly as is written in the applicable standards, they will provide a set of the similar results within the statistical bounds of the method. This way, it also helps to avoid unnecessary disputes and litigations, in addition to ensuring a uniform reporting of the test results by different parties executing the project, such as site engineers, quality control persons, and those involved in the third-party inspections. Because of this, the test data comprising different applicable sampling methods used to collect the materials, the purpose of testing, the methods and procedures, the prescribed evaluation criterion and the test results are an important part of any quality control and assurance policy of any project. These documents are assumed to be in possession of the concerned quality control engineers, and they will regularly update the same as and when new results pour in from the site. Based upon these, they will pass the necessary directions to the execution agency to rectify the defects if any come to their notice.

5.1.1 Data Generation

Data generation is a vital point in the material evaluation process. It is the process of creating a set of test data or test cases and the related applicable acceptance criterion required to check the adequacy and suitability of different materials for some construction project. It starts with the selection of some material attributes, based upon which the suitability of any material is to be decided. For instance, the dimensions of the bricks can be one of the deciding parameters to see their feasibility in the construction of a brick masonry building. The prescribed evaluation criterion for acceptance of clay burnt bricks is available in the BIS guideline, IS 1077-1992, which requires that twenty bricks selected in random from the sample should be placed along a line and their combined length must come within the prescribed range of 4600 $\pm$ 80 mm. A similar set of the range is given for the other dimensions of bricks. And, the minimum numbers of bricks required to constitute a representative sample of the bricks from the lot under considerations should be selected as per the mandate of the BIS guideline, IS 5454-1978. For example, it requires that a minimum of 40 bricks should be selected at random from a lot of 2001 to 10,000 bricks and any 20 should be taken at random from these 40 bricks to check their compliance with respect of the dimensions. The number of bricks will be more if the lot size is bigger than 10,000 bricks; depending upon the size of a lot, an appropriate number of bricks may be selected at random. Similarly, the acceptance criterion and the sampling guidelines are also prescribed for different materials by various regulatory bodies.

In India, the Bureau of Indian Standard (BIS) is one such body that formulates and regularly updates the applicable testing methods and procedures; design & construction specifications, and other guidelines related to the use of materials and the processes. For a data generation purpose, the selection of a suitable number of samples/specimens from the procurements or the different materials being required in various construction activities prior to their arrival at the site and documenting the related acceptance criterion is the very first key step in the process. Once these two parameters have been decided, the samples from the various lots/batches are collected in the required numbers for evaluating their response as per the applicable codal guidelines and the documented acceptance criteria and other key characteristics.

5.1.2 Data Collection

Data collection means obtaining necessary information from the selected sample of materials by means of appropriate testing. The test methods and related procedures are well-established and are documented in various guidelines. For instance, BIS has specified a number of test methods and related controls to achieve uniformity and repeatability (independent reproduction) in the test results. It is worth to note that these guidelines are dynamic in nature, and the regulatory body normally goes on updating the test rules over the time as and when some new information is consolidated based upon the research inputs and other performance studies. It is therefore utmost important to conduct the testing as per the provisions of the latest available codal provisions and guidelines. Testing a specimen can generate a lot of data, some of which may be of direct interest to the analyst/engineers in evaluating the materials under study, while the other may be of use for some other indirect applications. An attempt should be made to gather as much information from the test specimen as possible as it will save both the time and the cost associated with the preparation of new specimens at a later stage. It just needs only proper planning to execute the testing with the objectives primarily divided into two or more phases, namely the direct/primary set and the indirect/secondary set. The primary objectives are formulated by keeping in mind the short-term goals of the project, while the secondary ones are framed to meet the long-term goals. Some of the test data in the form of results are analyzed at the earliest possible, on priority, to see the outcome related to the primary objectives. The other part of the test result data can be taken in reserve to be analyzed at a later date, depending upon the secondary objectives of the testing and the project. For instance, we need to cast a certain number of concrete cubes to evaluate its mechanical properties. It involves materials, man-power as well as the very precious resource 'time'. The primary objective of the testing procedure can be just to see the compliance of the related BIS guidelines (IS 456-2000). So, we need to just check the highest compressive load that the standard cubes support before their failure. We will place the specimen in the loading machine and go on pressing it till it fails to hold any additional load. The load at the failure divided by the cross section of the specimen gives the compressive strength of the concrete. If the test results from the sample meet the acceptance criterion enshrined in the applicable codal guidelines e.g., IS 456-2000 in this case, the concrete is fit to be used in the construction activity. But, this loading of the cubes also generate a lot of other data points in terms of the displacements, both the longitudinal (along the loading axis) and in the transverse direction; their corresponding load values; weight of the cubes, both in the dry and the saturated conditions, etc. Thus, if we have set some monitoring points/devices to takes these measurements, this data can be analyzed easily to get additional material information, such as elastic modulus, the strain values (at the first crack and at the ultimate load, etc.), Poisson's ratio, and density, voids. This additional data can be a part of the secondary objectives of the testing/projects and is usually needed to evaluate the structural performance of concrete at the service and the ultimate states. Moreover, a set of the correlations can be developed between the in-situ concrete properties, such as its compressive strength, and density and ultrasonic pulse velocity, rebound number. It can be another subset of the secondary objectives. Such relations come very handy to quickly monitor the in-situ material properties at sites without the need of having any destructive testing. It did not involve anything extra, just needs framing of a proper testing strategy along with an appropriate testing sequence to get the relevant test data from a same set of the cube/sample to meet the identified objectives.

5.1.3 Compilation and Reporting

Compilation of the test data and its documentation is the final step of any testing assignment. It starts once the test results are obtained from appropriate testing methods. All decisions with regard to the suitability of materials and their response predictions depend upon this report. Therefore, it needs to be written with due care by experienced technician only and should be able to convey all information in a systematic manner, without bias, related to the observed laboratory response of the materials being reported. In general, the report should include a background summary, the methodology adopted to test the materials, and the test results. It must end by concluding the important findings, and based upon these, some recommendations can (should) also be made for the information of concerned stakeholders/users. If there is some limitation with respect to the testing procedure or the sample size used in the testing or any other observations, etc.; it is very important to highlight the same at some appropriate place in the report.

As all the results being reported depend upon the testing methods, equipment conditions, and the expertise of the testing team, it is good practice to include some brief information about them in the report. It should include what you have tested, where and when the test was conducted, equipment information, what you did during the test, details of your testing team members with regard to their expertise, capability, etc., and a brief account of the problems encountered as well as what worked well while performing the experiments.

Include all details of test methodology used in the experimentation so that others can recreate and reproduce the test results. In this section/paragraph of the report, an attempt should be made to explain how you have collected the sample of the materials and conducted the relevant tests by describing the number of specimens and their placement position, the monitoring points/devices used to take the measurements, the precautions exercised during the testing, and the metrics collected along with details of the equipment(s). The other details related to the testing standards used in the process such as the related BIS guideline also need to be included in the report. For instance, while reporting the test results of concrete cube, it is important to mention the relevant BIS guidelines, namely IS456 (2000), IS516 (1959) in the report as their control document. The former document defines the sampling and acceptance criterion while the latter gives details about the testing procedures. The details of the test data should be tabulated in the best possible way to convey the outcome of the testing. There is no fixed format for the tabulation; the number of rows and columns of the table can be selected appropriately to indicate the variables of the testing and their observed effects on each other. The rows and columns should be suitably named to clear the intent of the presentation.

In case of some negative observations, it is much more important to report and highlight any such test result if coming to the notice. It may be disheartening, but it helps the team and others to know where and what went wrong. Based upon this information, some suitable corrective action can be planned and executed to remove flaws if any in the production of materials or in the testing methods.

The report can be made more informative and interesting if some visual content appropriate to the subject matter is added. It can be photographic images, sketches, flowchart or block diagrams that help readers visualize what and how you were testing. An attempt should be made to add visuals in the report if something cannot be expressed suitably by means of text. These should be appropriately captioned and cited in the text to complement the subject matter being presented in the concerned paragraph. Short video clips or live images can also be included in the report to illustrate specific points about the observations or material response during the testing, if you are presenting the report electronically and the readers of the report have the technology available to see video clips. It will be very helpful to people who did not observe the actual testing sessions, but are interested to closely observe how the material being tested responded to the loading conditions.

And finally, list your findings and recommendations using all your data, both quantitative and qualitative. Each finding reported in the document should have a basis in the data—in what you actually observed and that noted during the testing operations/experimentation. The findings and recommendations can be given as just one overall list outlining the major outcomes; or it can be listed 'case-by-case' being reported in the document; or as a list of the major findings and recommendations that cut across all cases but giving their details as well as influence on each other. The negative observations if any, must be given due importance, and these should be mentioned suitably in this concluding section of a test report.

5.2 Method of Interpretation

Once the results are documented, the next keystone step in the material evaluation process is the interpretation of the reported test results. The results must be interpreted on the background of a certified reference that is used to distinguish between 'acceptable' and 'unacceptable'. The certified reference can be some of the applicable BIS guidelines/codes of practice or it can be a quality control/contract document mutually agreed by all concerned parties of the project. Depending upon the type of a material and the test parameters being evaluated, some suitable applicable guidelines/code of practice should be selected for taking a final call on the acceptability of materials for a construction project. While interfering, the engineers must be aware of the potential risk of false interpretation of the test results and its effect on the quality and costing of the project; it may result in increased costs on account of a frequent repair and maintenance that the effected components may require in future, during its service life and more importantly, the overall poor rating on public perception.

It is also of utmost importance that engineers should understand what is behind the test result with regard to statistical probability, random and systematic errors, etc. When a sample is tested several times or different specimens of a sample are tested, it is rare to get the same results every time. Instead, the test results will deviate more or less depending on the precision of the equipment and devices used in the testing, the controls being exercised during the production of materials along with the expertise of technician handling the equipments and also, how the measurements were taken during the testing operations. Thus, the two major contributions to the uncertainty are precision and accuracy, each contributing to random errors and systematic errors. Precision is the degree to which replicate measurements under the unchanged testing conditions show the same results while accuracy is the degree of closeness of measurements taken during the testing to its actual or accepted value. These two parameters must be paid due importance while interpreting the test results. The acceptable range is usually specified in the test reports supplied by laboratories. In case, it is not mentioned in the report alongside the reported test results, the applicable code of practice can be referred.

Table 5.1 Three-point scale to decide the acceptability of materials

Decision		Parameters to check while deciding & suggested corrective actions
Acceptable		The test results are well within their permitted range. Everything, including the material supply chain, production process, construction practice seems okay
Not acceptable	Critical	The results are outside the permitted range and that too, with a very large scatter. The material must be rejected or production/ construction processes needs to be modified on an immediate basis to meet the acceptance criterion
	Serious	The results are on margin, but a large scatter is observed in different lots of the materials. The production/construction processes need to be modified on urgent basis to meet the acceptance criterion
	Minor	The results are on margin. The supervision needs to be made better to bring the results within the prescribed limits

Source Author

Broadly, the material is either accepted or rejected depending upon the test results given in the report. However, if the results are found to be on the lower side of the acceptable limits; then, in such cases, an appropriate decision can be taken on a three-point scale, given in Table 5.1 to rectify the defects in the raw materials, or in the production process, or in the construction practice, or maybe at all these three listed levels.

5.3 Writing the Test Reports

Test reports are important project documents that store vital information about the materials used in the construction. All the major decisions related to procurement of different materials and their subsequent use in the work rest on these reports. It is therefore most important that these are written by experienced technician in a clear and concise manner without any ambiguity. While reporting the test results, it is necessary that the writer must include what is relevant to the problem under investigation. It is also important to mention the theoretical basis or the prior observations which have suggested the need of a particular experiment. The report should clearly present in reasonable detail the way the experiment was carried out, so as to permit someone else to duplicate the experiment and get a similar set of the results. At the end of the report, it must state what results were obtained and what interpretation of these can be made. Generally, every report should contain broadly the following sections:

Title: A good title indicates the relationships among the variables studied. It is a summary of the whole content of the report described in around 7–15 words. If it is an experimental study, uses verbs such as 'The effects of….on..' or 'The influence of…..on…' to reflect the causation. For correlation studies, refer to the direction of the relationships between variables, or type of the correlation being proposed, e.g., the possible title of such studies may be like this 'Correlation of Compressive Strength and Other Engineering Properties of High-Performance Steel Fiber-Reinforced Concrete.' The best way is to first write the content/main body of the report keeping in mind the research/testing objectives; Then, formulate conclusions based on the main outcomes of the study and then present the entire text in the form of one paragraph (i.e., Abstract). The title of the report/articles is then framed as a summary of the abstract; it will be a sentence describing the abstract of the report/article in one sentence of 7–15 words. It is therefore always better to write the title at the end of the work as it has to convey the intent of the entire document to the potential reader clearly and in an effective manner, which definitely will come from the abstract. It does not have any reference to the previous published work, or otherwise. The focus while writing the title should be the potential reader who must get a fair idea about the content of the article/report just by going through this one short sentence.

Abstract: It is a short summary (generally in about 250–550 words) of the entire test report/document/article that gives bird's overview about the statement of the problem, the methods used to obtain the reported results, and major outcomes that are subsequently presented in detail in the main body of the report. A brief and accurate summary of the full report allows the potential readers to determine quickly whether this study is relevant to their interests, and therefore, whether they should spend the time and effort to read it in full or not. Do not include any new material that has not been covered elsewhere in this report. Also, this is not the right place to present what was already available in the published literature in the field of the current study, which should be described briefly at the appropriate place in the main body of the report/article, preferably while introducing the problem statement.

Introduction: This section is an introductory statement of what led up to the experiment, and it must include as much background material from the published scientific literature as is needed to place the identified objectives of the study in its proper setting. While writing the introduction section of the report/experiment, try to answer these two questions: (a) what is known about this topic? and (b) Why was the experiment performed? It should be clear from the statements in the introduction what prior information is to be assumed in conducting the experiment(s) and where the sources of this information are to be found. It is likely that the experiment(s) may deal with some subject matters that are speculative, and these should be indicated and stated in terms that can reasonably be supposed to be generally and reliably understood. Few related past studies may be cited at the appropriate place in the text to justify the proposed speculations and prove that the proposed hypothesis is viable.

It should be kept in mind that the report is being written for everyone who might at any time, be interested in the said experiment(s) and not for one or two people who presumably already know all about the content. Since your report should be relatively 'timeless,' therefore the phrases such as '*a few weeks ago*' or '*last month*' should not be used. Information based on your personal experience should be avoided which others cannot evaluate or visualize. What the potential reader wishes to know is what contribution in terms of knowledge, concepts, test setups, available materials and methods and other precautions, etc., the report provides them. Some nobility of your work must be established in the background of previous available research studies. The introduction should end in a clear statement of the main aim of the experiment(s) being presented in the report.

Materials and Methods: This section of the report thoroughly describes how the experiment was carried out and must contain all details necessary to get the same or similar set of results if someone else is conducting the same experiment at some other location. In engineering, it is not sufficient merely to design and carry out an experiment. At the end of the day, others who are pursuing the similar interests/studies must be able to verify your findings, so your experiments have to be reproducible. The details given in the materials and methods section help them to follow the same procedure and obtain the similar results. Therefore, an attempt should be made to provide the details of what you did and, how you did it while performing the experiment. Be precise in providing details, but stay relevant. Ask yourself questions like this, 'Would it make any difference if this specimen were of different size/shape or made from some other material or produce or tested in some other way?' The answer may help you decide what to write and to what extent; and more importantly, to decide what to skip while writing the content of this section. Also, do not forget to provide the rationale behind your decisions in the selection of equipments, devices that were used to conduct the experiments and also, other information related to the sampling, specimens, production/casting methods, the testing methods, and the monitoring points. Explaining the rationale indicates that you understand the reasons for conducting the experiment in that way to meet the identified set of objectives, and that you are not just doing the things mechanically as the robots do.

Always express yourself in the past tense as you are providing a description of what you did in your experiment as it is already complete by the time you will start writing the report. The use of personal pronouns, such as *I* or *we*, must be avoided as it is against the very first principle of the scientific writing, which requires that the results being presented have to be reproducible under the reported laboratory conditions even if these are performed by some other competent person. However, using the first person in the writing indicates that it is not possible to obtain the similar results without the original researchers being present during the re-conduct of the listed experiments.

Sometimes, it becomes necessary to design and fabricates the testing equipments to meet the need to the experiment/hypothesis, especially when the available equipments/devices are not in a position to provide any help in meeting some of the identified objectives. In such cases, it becomes very important to provide the details of the equipment/experimental setup used in the study and the other aspects related to the procedure used in taking the measurements to the maximum possible extent to allow replication of the results. However, the details of the procedure/technique, or the instrumentation that have been adequately described by earlier investigators and are available in some published citable literature or if these are available as a standard testing procedure in the form of the standards and the codes/the BIS guidelines should not be described in details again, but references to the original sources should be cited.

Results: This section should present a description of the test data/test results collected during the experimentations and analyses performed on them. It is desirable to summarize the results in tabular or graphic form in addition to the verbal description in the report. A table or figure can frequently communicate your results far more effectively than words. Raw data, if any, should not be included in the results section; it is better to include this form of the data as an appendix to the report. Importantly, do not include interpretations of your data in the results section. Save that part of the report for its discussion section! Instead, simply present the data for what it is.

The figures, graphs, tables, flowcharts, block diagrams, etc., if any, being presented in the report should be concise and clear. They should be suitably captioned, numbered and referred to by that number in the text. The salient features observed from the test data presented in the tables, and graphs must also be described in the text of the results section. The results from statistical analysis of the test data, if any, should preferably be given at some appropriate paragraph in this section. However, if some comparison of the test data along with statistical details is made, it can be shown over the graphs also. Various word-processing software have a number of in-built features and functions for presenting the test data in different graphical forms. These should be explored and used to improve the presentation. However, it should be kept in mind that as of now, computer program cannot decide for you how your graph and table really works; you have to know how to design these presentation elements to convey the data in the best possible form to the potential readers.

Keep figures and tables as simple as possible; these are provided in the report/articles to dramatize the results in a manner that is easy to observe, compare and understand the trends for the potential readers. An attempt should be made to plot only three to five lines per graph; otherwise, a set of graphs can be used or a tabular format will be more appropriate to present a large data set.

It is a convention (and of course, it is also convenient) to read the data along the length of the pages from its top to bottom. So, the number of rows in tables should be kept more than the columns used therein, and the outcome of the experiment should be plotted on the vertical axis of the graph, which is along the length/height of the paper. During the experimentation, the independent variable is the condition that you have manipulated while the dependent variable is the outcome of that manipulation. The independent variable of the data therefore should be plotted on the horizontal axis of the graph and the dependent variable on its vertical axis. Do not forget to label the both axis of a graph or rows/columns of a table and be careful to include units of the measurements on the graph axis and, also in the tables.

Number and suitably title all graphs and tables being presented in the report. The titles should be informative and concise. These should be referred by the number in the text whenever is required to discuss the outcome of the study presented by means of these graphs or tables, e.g., 'Figure 1 shows the effect of the fiber aspect ratio on the workability of the concrete…….' Try to plot the graph as large as possible so that everything being presented through these is legible and clearly demarcated. Minimum 600 dpi over a width of 100 mm should be ensured for a better visibility in the prints.

Discussion: The purpose of this section is to provide an explanation of various trends and the patterns observed in the test results presented in the previous section of the report. An attempt must be made to cover the following points while writing the discussion section of the report:

- Relate the test results back to the objectives of the study and see whether these support your hypothesis or not.
- Compare the test results to the findings reported in other published scientific research to prove worthiness of the results or other test data being presented in the report.
- Provide an explanation of why the experiment produced the trends as are observed in the test results. Try to correlate the trends with some analytical model/theoretical basis or some physical mechanism that may have governed that set of the results.
- Identify problems in the test setup used in the study and suggest possible improvements to avoid such type of problems in future.
- State the practical and/or scientific significance of the test results.
- Present any anomalous data or any deviations from what you expected.
- Explore the theoretical and/or practical implications of your findings/reported outcomes.
- And finally, derive conclusions based on your findings.

After writing the results section, think carefully about how you can structure the discussion section in light of the above-mentioned points and achieve a logical flow of information from one topic to the next. In the discussion section, the results should not be simply restated; an attempt must be made to interpret the test results and derive some measurable outcomes.

References: References include all those published articles and other parts of the scientific literature that are relevant to the study and are used to build up the topic, identify objectives and validate the results or accuracy of the models being proposed in the report. It is the way of giving credit to all individuals whose intellectual works have been utilized to support your research/study. Typically, each such listed reference in this section includes the author's name, date and year of the publishing, journal title, the journal volume/issue number and the corresponding pages or digital object identifer (DOI) of the article. If some book, manual, conference proceeding, standard/codes have been referred, the details of the publisher and code/regulator may also be given for all such published articles and the papers referred in the report. A consistent style should be used in the citations. Any one of the available styles, such as APA, MLA or Chicago style can be selected. It is a norm to follow APA style for all articles belonging to education, psychology, technical articles while the MLA is mainly used by writers from the humanities background and the Chicago style is followed for reporting

Table 5.2 Citation style using the APA format

Type of the source	In-text citation	In the reference or bibliography list
A book	(Singh, 2017)	Singh, H. (2017). *Steel Fiber Reinforced Concrete: Behavior, Modeling and Design*. Springer, Switzerland
Chapter in an edited book	(Kaur et al., 2018)	Kaur, A., Singh, H., Jha, J. (2018). Experimental study of behavior of free-head flexible piles in two-layered sand under lateral and combined load. In H. Singh et al. (Eds.), *International conference on Sustainable Waste Management through Design* (1st ed., LNCE 21, pp. 116–124). Ludhiana (India)
An article in a print journal	(Singh, 2016)	Singh, H. (2016). "Strength and performance of steel fibre-reinforced concrete stiffened plates." *Magazine of Concrete Research*, 68(5), 250–259
An article in an electronic journal	(Singh, 2012)	Singh, H. (2012). Designing Laterally Loaded RC Stiffened-Plates: Energy-Balance Concept. *ISRN Civil Engineering*, *8*(3), 1–9. https://doi.org/10.5402/2012/709157
A website	(Flori, 2019)	Flori, M. (2019, November 20). *What Is Pervious Concrete*. New York Engineers. https://www.ny-engineers.com/blog/what-is-pervious-concrete

Source Author

the history, fine arts and business articles. Table 5.2 outlines some examples of referencing using the APA style.

There is a possibility that many articles that are not of any direct use are also used in the study. All such papers/articles should be listed in the bibliography section. Similar to the reference list, it may also contain journal papers; conference papers; design/testing manuals; books and codes of practice; online sources, etc., and these should be cited using the proper referencing style, but under a separate section at the end, named as a bibliography.

Some other important points that should be avoided while writing reports are listed below for the ready reference of a reader:

- Scientific writing should be clear, concise, objective and accurate.
- Use subheadings (if required) to make the flow of information systematic and clear.
- Check sentence structure. It should be simple and short and must be conveying what you are planning to convey and nothing else.
- Check each section flows logically from one paragraph to another.
- Use correct paragraph structure and try to present one topic per paragraph.
- Provides evidence and appropriate explanation about the topic being discussed.
- Concluding sentence should be clear outcomes of the study.
- Avoid using contracted words, e.g., won't, don't, might've, etc.
- Avoid using emotive language, e.g., beautifully described, looking heavy, etc.
- Avoid the use of personal pronouns, e.g., I, we.

- Use the past tense to describe what has been completed and the present tense to details the established facts. Make sure these are consistent throughout the text of the report/article.
- When beginning a new sentence with a number, write the number in words and not in the number itself, e.g., six cubes were used in this experiment, and not as 6 cubes were…
- Use the common names in the lower case and there is a no need to use italics.
- Introduce acronyms correctly, e.g., The American Concrete Institute (ACI) states that… It should be written in the bracket just after the full name being shown and it should be done at the very first paragraph of its use in the article/report.
- Be consistent with abbreviations and nomenclature throughout the text.
- At the end, be prepared to work on your report very carefully, with due patient, after you finished up the writing part. Editing your written work is very important and never forgets to proofread your article. Remember the simple fact that 'even after proof reading your article/report ten times, you will notice 10 points in your work even in the 11th attempt that needs one or other corrections to present the fact correctly.' So, be patient and go on reading and making corrections till you are able to convey what you are intending clearly and without any ambiguity.

5.4 Material Selection Process

Test reports contain information related to the material properties of different materials; the data related to their observed response may be under some mechanical or thermal loading conditions and an outcome in the form of conclusions and/or recommendations. The properties of materials and their response are the two vital parameters and prerequisites to any engineering design problem. Without these, we cannot estimate their weight—a source of all loads that we have to safely transfer to the ground and the material strength which we have to use to complete the load transfer process through various members of a building. The precise estimate of the weight of a material and its various strength parameters are the vital elements of any design process. Some materials are weighless that can be ignored in the calculations, while many others may have a much higher unit weight. At the same time, some may posses low strength values while others may be strong on one or more strength parameters. The selection of suitable materials to accomplish a safe transfer of the applied loads from the point of their applications into the ground and also, serve the other intended functions. It is, therefore, an important keystone in the construction cycle of any project.

A list of the materials available in the nearby vicinity of the project location is first prepared based upon the functional and structural requirements specified in the working drawings and the different specifications. The list is then examined to see

Table 5.3 Material attributes that needs to be evaluated while selecting them for some project

1	Availability in the local market and stocked in good quantities by relevant dealers
2	Availability of construction equipments needed in their production and handling
3	Constructability by means of the available skill set of the workers
4	Design life and compatibility of different materials if used together at the same place/member
5	Environmental issues, such as effects on human health, air, water, ground and animals
6	Listing by the relevant design code of practice for use within the stated design requirements
7	Past performance records and experience with their use in the similar environmental conditions
8	Standardization by recognized national and international standardization bodies
9	Thermal stability and aging resistance under the specified service temperature and other environmental exposure conditions.

Source Author

the feasibility of different materials for the intended use in the project under considerations, and the list is then further narrowed down, based upon the past experiences with respect to use of such materials and their past performance and other important attributes, given in Table 5.3. A survey form is often comes handy in this process. It can be generated for various materials needed in the construction using the attributes shown in Table 5.3 to make the process easy. Usually, the materials found suitable in the tentative final list are sampled from the place of their source and tested in the standard laboratory conditions to see their overall acceptability for the given field conditions and requirements of the construction project. Alternatively, the materials required for different components of the project can be selected, preliminarily, based upon the available laboratory test reports; often supplied by their manufacturers. However, this should be used only for the small size projects. And, for the larger ones, this practice should be avoided and an attempt should be made to conduct the actual testing on the material samples found suitable for the project. If the test results are found to be in the prescribed range of the acceptable limits, the material procurement process can be initiated by the concerned wing (often, procurement section) of the project. This is the very first step of quality control process.

The selection can also be made on the basis of the past experience where a similar type of materials is supplied and used regularly in the similar conditions. This usually applies to most of the materials that find routine applications in construction projects. However, a exception to this also exists when some new material is required for a special purpose or the available materials need to be used in some new applications and/or exposure conditions and/or certain dimensions, section sizes required as per the approved drawings are outside the range of the standardized available sections or outside the normal production range from manufacturers. Then in all such situations, the material selections must be made on the basis of relevant laboratory test reports produced by testing them under the anticipated field/site conditions. However, in such cases, some familiarization programs need to be conducted for the field staff and workers to get the desired level of quality and performance. The purpose of all such exercises should be to pre-qualify and verify the achievement of specified requirements on a consistent basis before the actual use of such materials in the construction activity starts.

On arrival of the material at the project site, a few tests are conducted (randomly or as per the terms of the contract document) to see the compliance of the purchase order and their general acceptability. If found okay, these are taken on record in the project inventory and later on, indented to the site staff/officials for further use. During the construction stage, the materials being used in the project are sampled as per the established procedures and tested in the laboratory under the standard conditions to see their compliance with respect to the project working drawings and the material specifications. In case of any deviation observed in the test results from the permitted controls, the required corrective actions become necessary to get the desired performance level. At all the levels of quality controls, the test reports are the backbone of the selection and quality assurance process and it needs to be documented as per the established procedures, either prescribed in the applicable code of practices or as per the provisions of the approved contract document. However, where the materials or fabrication or construction represent significant investments and/or operational and maintenance costs, a lifecycle cost analysis should be the basis of the material selection. Material selection in all such cases should be optimized, considering various investments and operational and maintenance costs, such that the lifecycle costs are minimized while providing an acceptable safety and reliability levels. This type of approach becomes mandatory over and above the routine work procedures described earlier just to make the project feasible over its service life span. The related documentation in this regard must include results from the relevant and independently verifiable tests, and/or confirmed successful experience under similar design, operational and environmental situations along with the results of lifecycle cost analysis conducted on the use of all such materials.

Bibliography

1. American Psychological Association (2020) Publication manual of the American Psychological Association: the official guide to APA style. Washington, DC
2. APA Blog (2020, August 15) APA style-grammar-guidelines/references. APA Style. https://apastyle.apa.org
3. Bary-Kahn P, Hildinger E, Hildinger E (2010) A practical guide to technical reports and presentations for scientists, engineers, and students. Pearson Custom Publishing, NY
4. Finkelstein L (Jr) (2004) Pocket book of technical writing for engineers and scientists. McGraw-Hill Higher Education
5. Laplante PA (2018) Technical writing: a practical guide for engineers, scientists, and nonotechnical professionals, 2nd edn. CRC Press, United Kingdom
6. Mcrobb M (2010) Specification writing and management. CRC Press, United Kingdom

7. Nelson JR (2003) Writing the technical report. McGraw-Hill Book Company, NY
8. O'Brien JJ (2013) Construction inspection handbook: quality assurance/quality control. Springer, US
9. Perry WE (1987) A structured approach to systems testing. Wiley, United Kingdom
10. Rumane AR (2013) Quality tools for managing construction projects. Taylor & Francis, United States
11. Rumane AR (2017).Quality management in construction projects. CRC Press, United Kingdom

Chapter 6
Code of Practice and Guidelines

Every profession is practiced by a number of people trained and skilled to work in their own chosen field. Everyone has their own level of expertise and vision to achieve perfection in the work they practice. Therefore, in order to provide a level playing field and fix a common minimum goal of the perfection, it becomes crucial for the regulators to document a set of the rules and regulations. All codes and standards of practice are formulated with this objective in mind, and these provide a guide to the knowledge, skills, judgment and attitudes that are needed for every professional to practice safely. These documents reflect a desired and achievable level of the performance against which the actual performance can be compared and documented. They are formulated primarily based on the established research and the performance studies after following a due consultation process and are approved for use by various professional bodies, such as the American Concrete Institute (ACI), the Japan Concrete Institute (JCI) or may be regulators, such as the Bureau of Indian Standards (BIS).

These documents generally complement the provisions of occupational health and safety laws and other regulations set by the governments to provide detailed practical guidance on how to comply with the legal obligations and simultaneously, tend to achieve a minimum standard of the desired quality. These documents are usually approved by various government bodies and carry the force of law, framed with a main objective to protect the public at large by setting up the minimum acceptable level of safety for buildings, materials and related processes. Thus, they have a special status being an approved document and are automatically admissible as evidence in court proceedings. A court may use these documents to establish what reasonably practicable action to manage a specific risk is and then, prove and fix the responsibility if something went wrong, knowingly or unknowingly. These documents are therefore an important cornerstone that provides a set of check and balances to achieve a desired minimum level of quality in the construction works by use of appropriate materials and practice and also, to fix responsibility whenever

H. Singh, *Structural Materials*, Materials Horizons: From Nature to Nanomaterials,
https://doi.org/10.1007/978-981-16-3211-2_6

something went wrong. They ensure a uniform engineering or technical criteria, methods and practice across the entire country for all professional and other stakeholders to follow. These documents form a set of 'accepted standards' in relation to the material specifications, the testing and other related information with respect to the materials testing and evaluation and also, of 'good practice' in relation to the design controls and construction procedures.

Normally, there are three types of documents, namely code of the standard, code of practice and model codes. A code of standard usually consists of technical definitions, procedures and/or guidelines that specify certain minimum requirements for engineers, manufacturers, firms/construction agencies and bidders to work for the achievement of some minimum level of the identified quality and safety. On the other hand, a code of practice is a standardized document that has been enacted into law by the government bodies having jurisdiction so that the engineers and other stakeholders, such as construction agencies, manufacturers are legally bound to comply with the provision of the applicable codes; non-compliance can result in being prosecuted. These documents include references to various standards; all such standards, therefore, automatically become part of the code and legally enforceable. The model codes, on the other hand, are not legally enforceable as these are developed by using the voluntary consensus standard process to compile and provide the latest research and technological knowhow and other related developments in the field of its application. Various professional bodies do this exercise and help the regulators to update their prevailing codes by customizing the model code as per the local needs. A model code is not enforceable until it is adopted by the government agencies. The material specifications and the related acceptance criterion should be frozen keeping in mind the provisions enshrined in these documents which have all important knowledge based on the research and practice available till the time of their release. In case of new materials and/or some emerging requirements of the projects, the related specialist literature and the results from performance studies should be used to frame the specifications, etc. While doing this, the priorities must be the expected performance and/or quality level of the to be constructed facility.

6.1 Quality Control and Importance

The quality has been defined as fitness for use; conformance to the requirements identified in the control documents, such as codes, standards or approved contract documents in the said activity/process and the pursuit of excellence. Quality control is a process used to fulfill the requirements for the quality, and it is a part of the larger quality management cycle, which thrives to achieve the identified and approved level of quality and safety, usually by means of a suitably formulated quality assurance process. The aim of carrying out the quality assurance is to prevent defects from entering into the work in the first place itself. Whereas the quality control process thrives to gauge and monitor the outcome of the work; it uses mutually agreed operational techniques and related activities as a reference to measure the quality of the ongoing work. The outcome of the whole exercise pays dividend in the form of client satisfaction, which will very likely result in both repeat business and referral business to the construction agencies, thereby the improved quality control so achieved became a competitive advantage for them. It also results in a reduction in the wastage of resources, financial gains and speedy completion of projects due to the reduction of funds spent on the rework tasks, scraps and related inventories. In case of poor quality control measures, the defect levels are found to be generally high that finally leads to a dent on the profitability of concerned construction agency in addition to reduced enthusiasm in the work force, unless steps are taken to tackle the root causes of the problems, and finally leads to enhanced project cost for the clients.

Quality control in construction works typically involves ensuring compliance with minimum standards/approved contract documents for both the materials and the workmanship in order to get the satisfactory performance in accordance with the approved design. For this purpose, random samples and statistical methods are commonly used as the basis for accepting or rejecting the work completed and the batches of materials being used in the construction works; the rejection simply means non-conformance of the relevant approved design specifications. The approved documents contain a clear-cut and unambiguous acceptance criterion for various materials to be used in the construction project and all other related design/construction parameters.

It is mandatory on part of all stakeholders of the project to ensure the compliance of approved documents in order to get the identified level of quality. It starts from the conceptual stage of the project; then, during its planning and design phase when different components/elements are proportioned and designed by a team of design engineers. Design reviews to ensure safe and effective construction procedures are a major element at this stage, which also controls the project schedule and its costing. Once the working drawings of the project are finalized, a detailed design, materials and construction practice-related specifications are written and/or identified from the standards. It may also include extensive training for the work staff on the use of specialized construction equipments, material handling or some new construction method. The possible sources of various types of materials being required and their potential suppliers are also given due considerations while freezing the design specifications, else the design is amended so as to make the construction feasible in the given set of constraints. For instance, the availability of good quality aggregates may be one of the major issues in the hilly terrain and if the proposed structural system needs high strength concrete (may be M40 or higher), then, in such cases, production of the concrete will be a major hindrance in the smooth progress of the work. And, ensuring that it happens as per the approved acceptance criterion will be a big challenge for the batching plant personnel. So, it is better to explore other structural materials in the design stage itself, may be mild steel rolled sections, but it again needs heavy construction equipments in their handling, to make the project workable and achievable in terms of the identified quality and to reduce avoidable litigations. The design team is required to explore all available options with regards to availability of the material, men, equipments and machinery using the 360 degree evaluation technique, wherein all possible options are studied from all angles to get the best solution.

An ideal quality control process requires the testing of all materials being deployed and the work being constructed. Testing so many samples normally leads to very high associated recurring costs and time; it also causes wastage of materials being tested. As a result, small size samples are commonly used to establish the basis of accepting or rejecting a particular work or material. Statistical methods are of great help to reach definite conclusions based on the test results regarding the acceptability of the materials and other constructed work in light of their approved specifications. These methods help to interpret the test results from a small size sample and derive conclusions applicable for the entire lot of materials. For instance, suppose that there are twenty undersized bricks in a lot of one hundred. In taking a sample of any five bricks, the quality control surveyor might not find any undersized brick or might have all sample bricks undersized. Drawing a direct inference that none or all bricks in the population are defective would be incorrect. Due to this random nature of the sample selection process, testing results can vary substantially. However, using statistical approach of interpretations tends to avoid such type of misleading conclusions from the small size of the sample. The detail of this approach is available in Chap. 4, which can be used to see the acceptability of different materials in light of the approved specifications either given in the contract document or control documents, such as codes and standards. The following sections in the chapter provide content that helps to choose a right material for a given set of the field constraints and freeze the related specifications.

6.2 Different Materials

This section presents specifications for some of the commonly used materials in the routine construction practice. These can be used to identify their acceptance based on the test results obtained from the samples of suitable size; the rejection of a lot of the material or some portion of the constructed facility using such type of materials that are found to defy the related design/material specifications.

6.2.1 Cementitious Materials

Cementitious materials are one of the key construction materials generally used to bind many other materials, mainly in the form of aggregates, together to get some new composite. The properties of the resultant material (e.g., mortar, concrete) differ significantly from their parent materials (e.g., sand, coarse aggregates). They usually contribute to the properties of the hardened concrete either through a hydraulic activity or pozzolanic activity, or both of these processes. Cement is one such type of material that finds numerous construction applications in the form of mortar and concrete. The advantage of cementitious materials -based composites is that they can be customized and designed easily to meet almost all forms of the design constraints and they allow the construction of different structural elements of complex geometrical shapes, unlike the metals that have very restricted such applications. The details and specifications of various such types of materials are listed below.

6.2.1.1 Lime

Lime is probably the oldest material that has been used since the ancient times to construct various buildings and other facilities. The Greeks and Romans had used this material extensively in the construction of many world famous buildings, and many of these are still standing today to exhibit the glory of the ancient construction techniques. The manner in which the structures built using lime -based materials has withstood the ravages of time, for about 2000 years old, bear irrefutable evidence to the durability of the material. It is produced when the limestone is heated to a temperature of around 850 °Celsius. The heating process, known as calcination, oxidizes the limestone to quicklime. The presence of other elements such as silica, alumina and iron in the impure limestone impart hydraulic characteristics to the resultant product after its calcination. This type of lime produced from the limestone is called as natural lime; whereas if it is produced by blending some pozzolanic/cements with the slaked lime, it is called as artificial lime. The process by which the limestone (calcium carbonate) is converted into the quicklime (Calcium oxide) by heating, then to the slaked lime (calcium hydroxide) by hydration, and which then, naturally reverts to calcium carbonate (original material in the form of limestone) by carbonation is called the 'lime cycle'; it is a natural process that occurs automatically depending upon the prevailing temperature and the moisture content. Therefore, it is the engineers who have to see which material in the cycle meets their requirements as a building material. The slaked lime is also called as hydrated lime; whereas in calcium oxide form, it is known as quicklime which is very reactive and undergoes an exothermic reaction in the presence of water/moisture. Traditionally, the calcination was carried out in small mixed-feed kilns which nowadays have been shifted to more refined and efficient shaft or rotary kilns.

Lime is mainly used as a mortar in the masonry work by mixing the lime and sand in suitable proportions. However, the use of lime as a plastering material is very limited owning to slow strength development and the long working time it needs to complete any given activity. The lime plaster hardens slowing by drying and recarbonation process. These are also not suitable for external surfaces in areas of the heavy rainfall, or very damp places and in the places where the water tightness is a major consideration. Moreover, metals coming in contact with the lime plaster are liable to corrosion; it is the biggest disadvantage of using lime as a structural material along with its low abrasive resistance and early hardness. However, these two ill-effects can be minimized by using coarser sand and fibers or by mixing some amount of cement (which itself is modified version of lime) in the lime plaster. Nevertheless, the lime plastered surfaces are especially suitable for receiving other finishes, such as white wash and other non-cement based materials because of their better adhesive properties in comparison with the other materials. So, these points should be given serious thought while freezing the specifications in comparison with the other competitive materials.

Lime is classified into six classes, namely A, B, C, D, E and F. Out of these, some types of the lime are hydraulic in nature, while others are non-hydraulic. Hydraulic lime is produced from impure limestone, which sets by undergoing hydrolysis in the presence of water; whereas the non-hydraulic lime hardens by reabsorbing the carbon dioxide from the air (this process is known as carbonation); it therefore takes much more time than the hydraulic lime to set and subsequently harden. The variations in the setting time define their applications in the routine construction activities; these are listed below:

Class A being eminently hydraulic in nature is most suitable for the structural applications, especially in the outdoor/exposed surfaces, such as paving, walls, and chimneys.

Class B is semi-hydraulic in nature and is mainly used in the masonry mortars, lime concrete, and also in plastering work.

Classes C, D and F find their use as a finishing coat in the plastering and white-washing work. However, blending these types of the lime with some pozzolanic materials makes it fit for use in the masonry mortar.

Class E, also known as Kankar lime, used extensively as masonry mortars.

The percentage of active clay material (alumina and silica) in the lime decides its quality and reactivity, and the consequential hardening process that occurs in the presence of water. The mineral content less than 12% (known as feebly hydraulic lime, mostly belongs to classes C to F) makes the material relatively slow setting and bonding remains poor in comparison with the lime having this content in the range of 12–18% (known as moderately hydraulic limes; class B) or a higher range of 18–25% (known as eminently hydraulic limes; class A). Therefore, a batch of lime will be acceptable for use in the construction only if it meets the minimum chemical composition and physical requirements. It is given in Tables 6.1 and 6.2. It shall be tested as per the provisions of IS 1514 [4]. While taking a sample from the given batch/lot, all the containers in a single consignment of the lime drawn from a single manufacturing batch shall be considered, which constitute the given lot. If a consignment is declared or known to consist of different batches of manufacture, the batches shall be marked separately and the containers belonging to the same batch shall be grouped together and each such group shall constitute a separate lot of material for the testing purpose.

Table 6.1 Chemical composition of various forms of lime

Lime class		A	B		C		D		E	F	
		Hydrate	Hydrate	Quick	Hydrate	Quick	Hydrate	Quick	Hydrate	Hydrate	Quick
Calcium and magnesium oxides (%) Min. on ignited basis		60	70	70	85	85	85	85	50	70	70
Magnesium oxides (%) on ignited basis	Max	6	6	6	6	6	–	–	6	–	–
	Min	–	–	–	–	–	6	6	–	6	6
Silica alumina and ferric oxide (%) Max. on ignited basis		20	10	10	–	–	–	–	20	10	10
Unhydrated magnesium (%) Max. on ignited basis		–	–	–	–	–	8	8	–	8	8
Carbon dioxide (%) Max. on oven dry basis		5	5	5	5	5	5	5	5	5	5
Available lime as CaO (%) Min. (on dry basis)		–	–	–	75	75	–	–	–	–	–

– Denotes non-specified entries
Source https://archive.org/details/gov.in.is.712.1984

Table 6.2 Physical requirements for the lime

Lime class		A	B		C		D		E	F	
		Hydrate	Hydrate	Quick	Hydrate	Quick	Hydrate	Quick	Hydrate	Hydrate	Quick
Setting time (min)	Initial	2	–	–	–	–	–	–	2	–	–
	Final	48	–	–	–	–	–	–	48	–	–
Compressive strength (MPa)	At 14 days	1.75	1.25	1.25	–	–	–	–	1.00	1.25	1.25
	At 28 days	2.80	1.75	1.75	–	–	–	–	1.75	1.75	1.75
Soundness Chatelier expansion (%)		5	–	5	–	–	–	–	10	–	10

– Denotes non-specified entries
Source https://archive.org/details/gov.in.is.712.1984

It is worth to note that the lime belonging to the classes C and D is mostly used as a white wash material so there is no need of any setting time requirements or for it to exhibit compressive strength in various structural applications.

6.2.1.2 Cements

Cements are the man-made materials which posses binding characteristics just similar to the lime, albeit much stronger than it. Basically, cements are the modified form of lime, produced in mechanized plants under a controlled set of the conditions. Cements are produced in four stages, namely crushing and grinding of raw materials, mainly calcareous raw materials (limestone, chalk and coral deposits) along with argillaceous raw materials (Clay, shale and slate); blending the materials in the correct proportions; burning the prepared mix in a kiln at the temperature of 1350–1550 °Celsius; and finally, grinding the burned product, known as 'clinker'. At this stage, the gypsum is usually added to control the setting characteristics of the cement. Unlike the lime, the cement consists of compounds of lime (calcium oxide, CaO) mixed with the silica (silicon dioxide, SiO_2) and the alumina (aluminum oxide, Al_2O_3). The calcareous raw materials supply the lime in the cement powder; whereas the other oxides required to meet the desired chemical compositions in the cement are obtained from the argillaceous material. Additional raw materials such as silica sand, iron oxide and bauxite are also added in small quantities to get the desired composition. Industrial wastes, such as fly ash and calcium carbonate are other possible raw materials, but their use is relatively small in comparison with that of the natural materials, but this addition though in small quantities helps to reuse the industrial by-products in a systematic manner. The Portland pozzolanic cements (PPC) are the special type of this group that is blended with fly ash at the time of its production and helps to optimize the use of limestone in the production process.

The cement in the final ready-to-use powder form consists of four main compounds, namely tricalcium silicate ($3CaO{\cdot}SiO_2$), dicalcium silicate ($2CaO{\cdot}SiO_2$), tricalcium aluminate ($3CaO{\cdot}Al_2O_3$) and tetra-calcium aluminoferrite ($4CaO{\cdot}Al_2O_3Fe_2O_3$). These compounds are abbreviated as C3S, C2S, C3A and C4AF, respectively, where C stands for calcium oxide, S for silica, A for alumina and F for iron oxide present in the cement powder. The strength development in the cement depends on its chemical composition and the fineness to which it is grounded; the finer the cement particles the more rapid it would develop the strength. Depending upon the fineness of cement particles obtained in the grinding process, three types of cements are in use in India, namely C33, C43 and C53. Out of these, C53 is the finest and it develops strength more rapidly than the other two grades. Basically, it exhibits early strength owing to the presence of the C3S while the C2S contributes toward the subsequent increase in strength with time. The fineness of cement particles also helps to accelerate the hydration process. The other compounds such as that of alumina and iron make a little direct contribution to the strength but help otherwise to control some other properties of the cement. Their relative ratio in the cement plays a key role in defining the strength and other properties.

The guidelines and other codes put some constraints on these ratios, so as to ensure the achievement of certain minimum strength values. Separate codes are formulated for different types of cements, e.g., ordinary Portland cement for general use, low heat Portland cement for use in the areas where low heat of hydration is required, such as mass concreting works; the rapid-hardening Portland cement where high-early strength is required, etc. Use of the sulfate resisting Portland cement is particularly beneficial in all such conditions where the concrete is exposed to the risk of deterioration due to sulfate attack, e.g., in the members coming in direct contact with the soils and groundwaters containing excessive amounts of sulfates; it is done by limiting the C3A and C4AF in the mix. On the other hand, the high alumina cement is mainly used in refractory applications and also in some cold regions where the high heat of hydration and high-early strength development gives an added advantage in such cold temperature conditions. It also finds use in work involving emergency repair or construction. However, the reduction in the strength due to increased porosity in the matrix may be kept in mind while deciding its use as the water released during the hydration reaction leads to more voids in the matrix. Similarly, the supersulfated cement is best suited in a variety of the aggressive conditions, e.g., for marine works, concrete construction in sulfate bearing soils or liable to the attack by aggressive water, and in chemical works under conditions involving exposure to high concentrations of sulfates of weak solutions of mineral acids, concrete sewers carrying industrial effluents. Variations in the physical characteristics in different cements are obtained by changing their chemical composition, as given in Table 6.3; the resultant minimum set of the physical properties is given in Table 6.4. The cements should be tested in the laboratory conditions to see the compliance of these listed composition and other physical properties; non-compliance must lead to the rejection as the strength, and other performance parameters of all built structures depend only on the positive results.

The cement before the use must feel cool on touching, and there should not be any dry lumps in the powder. Additionally, cement may be rejected if the test results do not comply with any of the requirements summarized in Tables 6.3 and 6.4. Moreover, if the cement that remains in bulk storage at the factory or at the dump for more than six months or the cement in bags, in local storage such as, in the hands of a vendor or site for more than three months, should be retested to see the compliance and shall be rejected if it fails to conform to any of the listed requirements of their respective standard.

6.2.1.3 Pozzolanic Materials

A pozzolanic material consists of siliceous or a combination of siliceous and aluminous materials in their finely divided form. They themselves posses zero or negligible binding properties but on coming in contact with moisture, start

Table 6.3 Chemical requirements for various types of cements

S. No	Types of cement	Ratio of percentage of lime to percentages of silica, alumina and iron oxide	Ratio of percentage of alumina to that of iron oxide (%) (Min.)	Insoluble residue, percent by mass (Max.) * *X is declared percent pozzolana*	Magnesia, percent by mass (Max.)	Total sulfur content calculated at sulfuric anhydride (SO_3), percent by mass (Max.)	Loss on ignition (%) Max	Chloride content, percent by mass (Max.)
1	33 Grade OPC (IS 269–1989)	0.66–1.02	0.66	5	6	3.5	5	0.1
2	43 Grade OPC (IS 8112–1989)	0.66–1.02	0.66	4	6	3.5	5	0.1
3	53 Grade OPC (IS 12269–1987)	0.80–1.02	0.66	4	6	3.5	4	0.1
4	Sulfate resisting cement (IS 12330–1988)	0.66–1.02	–	4	6	2.5	5	–
5	Portland pozzolana cement—Part 1 (IS 1489–1991)	–	–	$X + 0.04(100 - X)$	6	3.0	5	–
6	Rapid-hardening cement (IS 8041–1990)	0.66–1.02	0.66	4	6	2.5 for C3A < 5 and 3 for C3A > 5	5	–
7	Portland slag cement (IS 455–1989)	–	–	4	10	3.0	5	–
8	High alumina cement (IS 6452–1989)	–	–	–	–	–	–	–

(continued)

Table 6.3 (continued)

S. No	Types of cement	Ratio of percentage of lime to percentages of silica, alumina and iron oxide	Ratio of percentage of alumina to that of iron oxide (%) (Min.)	Insoluble residue, percent by mass (Max.) * *X is declared percent pozzolana*	Magnesia, percent by mass (Max.)	Total sulfur content calculated at sulfuric anhydride (SO_3), percent by mass (Max.)	Loss on ignition (%) Max	Chloride content, percent by mass (Max.)
9	Supersulfated cement (IS 6909–1990)	–	–	4	10	6.0	–	–
10	Low heatportland cement (IS 12600–1987)	–	0.66	4	6	2.5 for C3A < 5 and 3 for C3A > 5	5	–
11	Cement 53-S (12,269–1987)	0.66–1.02	0.66	3	5	3.0	4	–

Source https://archive.org/details/gov.in.is (respective codes, as given above)

Table 6.4 Physical requirements for various types of cements

S. No	Type of cement (standard)	Fineness (m^2/kg) (Min.)	Soundness (percent)		Setting time (minutes)		Min. compressive strength (MPa)			
			Le Chatelier (Max.)	Autoclave (Max.)	Initial (Min.)	Final (Min.)	1 Days	3 Days	7 Days	28 Day
1	33 Grade OPC (IS 269–1989)	225	10	0.8	30	600	–	16	22	33
2	43 Grade OPC (IS 8112–1989	225	10	0.8	30	600	–	23	33	43
3	53 Grade OPC (IS 12269–1987)	225	10	0.8	30	600	16	27	37	53
4	Portland pozzolana cement—Part 1 (IS 1489–1991)	300	10	0.8	30	600	–	16	22	33
5	Sulfate resisting cement (IS 12330–1988)	225	10	0.8	30	600	–	10	16	33
6	Low heatportland cement (IS 12600–1987)	325	10	0.8	60	600	–	10	16	35
7	Rapid-hardening cement (IS 8041–1990)	325	10	0.8	30	600	–	27	–	–
8	Portland slag cement (IS 455–1989)	225	10	0.8	30	600	–	16	22	33
9	High alumina cement (IS 6452–1989)	225	5 mm	–	30	600	30	35	–	–
10	Supersulfated cement (IS 6909–1990)	400	5 mm	–	30	600	–	15	22	30
11	Masonry cement (IS 3466–1987)	–	10	1	90	1440	–	–	2.5	5

Source https://archive.org/details/gov.in.is (respective codes, as given above)

exhibiting cementitious properties. At the ordinary temperatures, they react with the calcium hydroxide or calcium oxide coming into the solution to form the compounds that possess cementitious properties. The capacity of such materials to react with calcium hydroxide in the presence of water is a measure of their pozzolanic activity; the higher the value the better cementitious characteristic they would have. This process is irreversible, and it usually occurs in the highly alkaline environment ($pH > 12$) that ensures good solubility of silicon and aluminum ions to support the pozzolanic reaction. As a result, the pozzolanic materials are in use since ancient times as a construction material along with the lime; their mixture in certain proportions imparts them cementitious properties needed to construct the buildings and other components.

The rate of the pozzolanic reaction depends upon the mix proportions, the amount of water or empty space available in the matrix to permit the formation and growth of hydration products and the temperature of reaction. The products of the usual hydration reaction and the high prevailing alkalinity in the cement impart the desirable environment required to sustain the pozzolanic reaction. Consequentially, these are used extensively in the production of blended cements, such as the pozzolanic Portland cement (PPC).

Pozzolanic materials occur naturally as well as are available as the by-products from many manufacturing plants. They are found naturally as volcanic ash, tuffs, pumicites, clay and shale. Natural pozzolans need further grinding to convert them to a finely divided form and then, calcining to activate them as a pozzolanic material. Alternatively, they can be produced artificially by thermal activation of the kaolin clays to obtain metakaolin, which is a good pozzolanic material. They are also obtained as industrial by-products from the high-temperature process, like those used in the thermal power plants. Fly ash and silica fume are examples of the pozzolanic materials that are obtained from various industries as their waste, but have a significant value otherwise in the cement industry. These materials are suitable as a building material only if they possess a high pozzolanic activity and that helps them to modify the microstructure of the resultant material. Various guidelines and the applicable codes attempt to fix certain minimum parameters that all pozzolana must meet to qualify themselves as good cementitious materials.

Fly ash: It is an industrial by-product produced during the combustion of pulverized coal and collected by electrostatic precipitators in the exhausts of thermal plants. Most of the volatile matter and carbon present in the coal are burned off during the combustion and other mineral impurities, such as clay, quartz and shale present in the coal fuse in suspension, wherein they rapidly solidify and lead to the formation of fly ash particles. The major consequence of the rapid cooling is that few minerals have time to crystallize, while others remain amorphous in nature. The particles are generally spherical in shape and range in size from 1 to 300 μm. The mineralogy of fly ashes is very diverse; they contain a heterogeneous mix of various compounds, such as SiO_2, Al_2O_3, Fe_2O_3 and CaO in varying proportions, depending upon the fuel used in the heating processes. The surface area of fly ash particles typically varies from 300 to 500 m^2/kg; although in some types, it is found to be as low as 200 m^2/kg and in others as high as 700 m^2/kg. The bulk density varies from 550 to 850 kg/m^3 in the loose state, whereas in the dense state, it ranges between 1100 and 1500 kg/m^3 depending upon the level of compaction. Table 6.5 lists typical chemical compositions of the fly ash generated in the thermal plants from the use of various types of coals as their fuel.

Depending upon their chemical composition, fly ash is classified into two classes, namely calcareous fly ash (also known as class C) and siliceous fly ash (also known as class F). The class C fly ash contains reactive calcium oxide in quantities not less than 10% by mass and because of this; it possesses good pozzolanic and hydraulic properties. Burning of sub-bituminous or lignite coal produces this type of the fly ash. Otherwise, the fly ash is classified as Class F. It is produced from burning of anthracite and bituminous coal and contains more silica and iron content than the (much needed) lime (CaO) content to initiate the pozzolanic reactions. Such types of fly ashes acquire the cementitious properties when some cementing agents, such as Portland cement, quick/hydrated lime are added to the fly ash. The fly ash must meet the chemical requirements given in Table 6.6 to be fit for use in the cement production as well as other construction activities.

In addition, for using fly ash as a part of the pozzolanic Portland cement, the fly ash particles found retained on 45 micron sieve should not be more than 34%; neither should it have a specific surface area of value not less than 320 m^2/kg, nor a

Table 6.5 General chemical composition of commonly found fly ash from different types of coals

Compounds (in percent)	Source of fly ash		
	Anthracite and bituminous coal	Sub-bituminous coal	Lignite coal
SiO_2	20–60	40–60	15–45
Al_2O_3	5–40	20–30	20–25
Fe_2O_3	10–40	4–10	4–15
CaO	1–10	5–30	15–40
Lose on ignition	0–15	0–3	0–5

Source Author

Table 6.6 Mandatory chemical composition of the fly ash for use in cements (IS 3812)

Parameters (in percent by mass)	Class F	Class C
$SiO_2 + Al_2O_3 + Fe_2O_3$, Minimum	70	50
SiO_2, Minimum	35	25
Reactive silica, Minimum	20	20
MgO, Maximum	5	5
Total sulfur as sulfur trioxide (SO_3), Maximum	3	3
Available alkalis as equivalent sodium oxide (Na_2O)	1.5	1.5
Total chlorides, Maximum	0.05	0.05
Loss on ignition, Maximum	5	5

Source https://archive.org/details/gov.in.is.3812.2013

lime reactivity of less than 4.5 MPa (measured in terms of compressive strength). However, if the fly ash is to be used as admixtures in cement mortar and concrete, these values must meet requirements of retention on a 45 micron sieve of not more than 50% and the specific surface area of not less than 200 m^2/kg. In the either case, the individual results for these parameters should not vary more than 15% from the average value established from the tests on the ten material samples; otherwise, the sample should be rejected.

Silica Fume: It is very fine pozzolanic material; it forms as an industrial by-product during the production of element silicon or ferrosilicon alloys due to the reduction of high-purity quartz with coal used in electric arc furnace. Silica fume rises as an oxidized vapor during the process and when it cools, the particles condenses and is collected in huge cloth bags as a very fine powdery material. It mainly composed of silicon dioxide in the amorphous form, mainly spherically shaped particles, having a diameter of less than 1 μm and with an average diameter of about 0.1 μm. These particles are about 100 times finer than the average cement particles and possess a very large surface area in the range of 10,000–20,000 m^2/kg. The bulk density of the silica fume varies from 125 to 450 kg/m^3.

For using the Silica fume as a mineral admixture in various construction-related activities, it must have silica content (SiO_2) of more than 85% (by mass). Alkalis as Na_2O if present should not be more than 1.5%; neither lose on ignition should be more than 4%. The minimum specific surface area of the silica fume particles must be 15,000 m^2/kg and the oversized particles retained in 45 micron sieve should be limited to 10% (by mass).

Slag: It is an industrial waste left over after a desired metal has been separated from its raw ore. Slag is usually a mixture of the metal oxides and silicon dioxide; it may also contain calcium oxides depending upon the type of ore and the extraction process used to get the metal. For instance, the molten slag from the iron blast furnace if rapidly chilled by quenching in water leads to the formation of a glassy sand-like granulated material, known as ground granulated blast furnace (GGBF) slag. It contains various compounds, namely CaO (30–50%), SiO_2 (25–40%), Al_2O_3 (5–25%) and MgO (1–20%). Because of the presence of CaO and SiO_2 in the mix, the GGBF slag also exhibits pozzolanic properties in the presence of water. It is worth to note that air-cooled slag does not have the hydraulic properties, as are exhibited by the water cooled slag. The granulated slag is usually grounded to a size of about 45 μm or less, and it must possess a specific surface area of about 400–600 m^2/kg along with a bulk density of 1050–1375 kg/m^3.

GGBF slag is commonly used to produce sulfate-resistant cements, called as Portland slag cement. The rough and angular-shaped ground slag particles easily hydrate in the presence of water and CaOH supplied by the cement during the hydration process. It sets in a manner similar to the ordinary Portland cement and possesses better sulfate resistant characteristics than the conventional cements produced for this purpose. The slag is considered fit for the blending with the cement clinkers if it meets the requirements given in Table 6.7.

The slag content is limited between 25 and 65% of the cement content to get the desirable characteristics in the final product.

Table 6.7 Permissible chemical composition of slag for use in the cement production

Composition	Maximum permitted value (in percent)
Magnesium oxide (MaO)	8·0
Sulfur trioxide (SO_3)	3·0
Sulfide sulfur (S)	1·5
Loss on ignition	5.0
Insoluble residue	4.0

Source https://archive.org/details/gov.in.is.455.1995

6.2.2 Aggregates

Aggregates are the most mined materials in the world, which constitute a broad category of particulate material, usually consisting of fine- to coarse-sized particles. It includes particles in the form of sand, gravel and the stone crushed to various sizes. The aggregates find enormous applications in road/railway track construction, concrete production and as a filter media. It is mainly classified into two broad classes, namely fine aggregates and coarse aggregates depending upon the size of particles it composed of. Each class of the aggregates has their role to play as a building material. In some activities, such as plastering and wearing coats in the road construction only the fine aggregates mixed with cement or bitumen, respectively, are used; whereas in some other applications, such as activities related to construction of roads/railway tracks, these are used in coarse form; while in the concrete production, we need both the fine and the coarse aggregates to form bulk of the concrete volume. But each application needs aggregates conforming to the task specific specifications; it is described below.

6.2.2.1 Fine Aggregates

The aggregate most of which passes 4.75 mm IS Sieve is classified as fine aggregates. It can be produced artificially by crushing stones and gravels and it also exists naturally in the river beds where the stones/gravels disintegrate and erode to the particle sizes making them fit to classify as fine aggregates. Depending upon the mode how these have been formed, these are called as crushed stone sand or the natural sand, respectively. Irrespective of their source, they should be hard, strong, dense, durable, clear and free from alkali, vegetable matter and other deleterious substances; Table 6.8 indicates the permitted values of these deleterious materials in the fine aggregates that are permitted to be used in the concrete production.

The fine aggregate is further divided into four grading zones depending upon the range of different particle sizes, they possess, namely grading zones I, II, III and IV. The distinction is made on the basis of the percent passing through 600 μm IS

Table 6.8 Limits of deleterious materials in fine aggregates for their safe use in the concrete production

Deleterious substance	Uncrushed fine aggregate (percentage by mass, Max.)	Crushed/ blended fine aggregate (percentage by mass, Max.)
Coal and lignite	1	1
Clay lumps	1	1
Materials finer than 75 μm IS Sieve	3	10
Soft fragments	–	–
Shale	1	–
Total of percentages of all above deleterious materials (except mica)	5	2

Source https://archive.org/details/gov.in.is.383.2016

Table 6.9 Grading zones for fine aggregates for use in the concrete production

IS sieve designation	Percent passing for different grading zones			
	Zone-I	Zone-II	Zone-III	Zone-IV
10 mm	100	100	100	100
4.75 mm	90–100	90–100	90–100	95–100
2.36 mm	60–95	75–100	85–100	95–100
1.18 mm	30–70	55–90	75–100	90–100
600 μm	15–34	35–59	60–79	80–100
300 μm	5–20	8–30	12–40	15–50
150 μm	0–10	0–10	0–10	0–15

Source https://archive.org/details/gov.in.is.383.2016

sieve. The grading limits for the different zones of fine aggregates are given in Table 6.9.

In cases where the grading falls outside the limits of any particular grading zone of sieves, other than 600 μm IS Sieve, by a total amount not exceeding 5%, it shall be regarded as falling within that grading zone. For the sand produced from the crushed stone, however, the tabulated permissible limit on 150 μm IS Sieve is permitted to increase by 20% and up to 5% for the other sieve sizes. The sand belonging to zone-I and II is not permitted for use in the concrete production under normal circumstances, until and unless the necessary corrections are made in the mix proportions and/or the sand belong to other groups is not locally or economically available. However, the sand from these two groups can be used in the plastering work and other similar activities.

6.2.2.2 Coarse Aggregates

Aggregates most of which are retained on 4.75 mm IS Sieve and contain only the permitted quantity of the fine content in the mix are called coarse aggregates. Like the fine aggregates, irrespective of their source, these are also classified into various sizes and they should be hard; strong; dense; durable; and free from alkali, vegetable matter and other deleterious substances. If some deleterious materials are present, their content should be limited to the values listed in Table 6.10.

If the deleterious content in the aggregates is found to be well within the prescribed limits, these are sieved to identify their nominal size. It is defined by the sieve size where the quantities of the material passing are found to be in the range of 85–100%. Usually, this value is used in the design of concrete mix proportions. The percent passing through different sieves are given in Table 6.11. The given sample of coarse aggregates is sieved, and accordingly, these are classified depending upon the percent passing.

Aggregates find their use in a verity of different construction works, wherein they transfer the load mainly by means of the surface contact that they would have among each other. During this process, they have to bear the stresses caused by the load through whatever contact area they would have with the adjoining aggregates. Use of a mix of fine and coarse aggregates in certain proportions tends to improve the contact area, and they perform better than the single-sized aggregates, albeit in compression only. But, the addition of some cementitious materials in the mix improves the response even under the tensile loading, e.g., the concrete possesses better load carrying capacity in tension than a pack of the aggregates otherwise.

There are many other mechanical properties also that the aggregates have to meet to perform their intended function satisfactorily. The aggregate crushing and impact values are important parameters between them. The aggregate crushing value and their impact value should not exceed 30% whenever they are used in concrete for different wearing surfaces, such as roads, runways, pavements, spillways and stilling basins. And for the aggregates to be used in concrete other than for the wearing surfaces, both of these can be allowed up to 45%. The tests should be conducted as per the provisions of IS 2386. Basically, these values indicate the

Table 6.10 Limits of deleterious materials (percent mass, Max.) in coarse aggregates

Deleterious substance	Uncrushed coarse aggregate	Crushed/blended coarse aggregate
Coal and lignite	1	1
Clay lumps	1	1
Materials finer than 75 μm IS Sieve	3	3
Soft fragments	3	–
Total of percentages of all deleterious materials (except mica)	5	5

Source https://archive.org/details/gov.in.is.383.2016

Table 6.11 Nominal size of coarse aggregates depending upon their grading

IS Sieve designation	Percentage passing for single-sized coarse aggregate of nominal size						Percentage passing for graded aggregate of nominal size			
	63 mm	40 mm	20 mm	16 mm	12.5 mm	10 mm	40 mm	20 mm	16 mm	12.5 mm
80 mm	100	–	–	–	–	–	100	–	–	–
63 mm	85–100	100	–	–	–	–	–	–	–	–
40 mm	0–30	85–100	100	–	–	–	90–100	100	–	–
20 mm	0–5	0–20	85–100	100	–	–	30–70	90–100	100	100
16 mm	–	–	–	85–100	100	–	–	–	90–100	–
12.5 mm	–	–	–	–	85–100	100	–	–	–	90–100
10 mm	0–5	0–5	0–20	0–30	0–45	85–100	10–35	25–55	30–70	40–85
4.75 mm	–	–	0–5	0–5	0–10	0–20	0–5	0–10	0–10	0–10
2.36 mm	–	–	–	–	–	0–5	–	–	–	–

Source https://archive.org/details/gov.in.is.383.2016

extent to which they would crush/disintegrate into the specified finer fractions under the action of applied stresses; the higher the value, the more they will crush into the finer fraction. Similarly, the aggregate abrasion value should be restricted to 30% for concrete to be used in the wearing surfaces, and for the non-wearing surfaces, it is permitted up to 50%.

The stability of aggregates under aggressive environmental conditions, such as exposure to certain chemicals, is a measure of their soundness. These chemicals may come in contact with the aggregates from the use in such conditions or from the water/moisture that may bring them in contact with the chemicals. To be fit for use in concrete, the average loss of mass after 5 cycles shall not exceed 10 and 12% when the fine and coarse aggregates, respectively, are tested with sodium sulfate (Na_2SO_4) solution; the test should be performed as per the provisions of IS 2386 (Part 5). In some cases, they may be generated within the concrete as a product of hydration reactions in the form of CaOH, NaOH, etc. They normally lead to cracking within the materials because of the expansive nature of the reactions that takes place between the alkali reactive aggregates and the prevailing alkalinity in the solution. To decide their suitability under such conditions, the mortar bar test should be conducted as per the provisions of IS 2386 (Part 7); the permissible limits for the mortar bar test at 38 °C shall be 0.05% 90 days and 0.10% at 180 days. In case of slowly reactive aggregates, these limits are prescribed as 0.05% at 90 days and 0.06% at 180 days for the test conducted at 60 °C.

6.2.3 Admixtures

The materials other than aggregates, water or cementitious materials (such as cement, fibers or pozzolana) that are used as an ingredient of the concrete or mortar and added to the batch immediately before or during its mixing to modify one or more properties in the plastic or hardened state is known as admixture. There are basically four types of admixtures, namely accelerating admixtures, retarding admixtures, water-reducing admixtures, air-entraining admixtures and superplasticizing admixtures. Each one has a unique purpose and modifies one or more properties of the concrete or mortar. For instance, an admixture added to concrete, mortar or grout with a purpose to increase the rate of hydration of the cement, thereby shortening its setting time is called as accelerating admixtures, while the retarders are used to delay these parameters. On the hand, the superplasticizers are added in mortar or concrete to increase their workability without increasing the water content or maintain workability with a reduced amount of the water. The air-entraining admixtures are added to introduce tiny air bubbles in concrete or mortar during mixing to increase workability and impart freezing and thawing resistance to the material. As these compounds have a significant effect on the concrete/mortar mechanical properties, we have to be very careful in the selection of their dosage and the type required to achieve the desired objective. Therefore, the selection process requires a number of trials to decide which type and admixture dose will not adversely affects the concrete strength and its other properties. Whatever is finalized at the time of the concrete mix designing, it should be ensured to use same without any plus-minus in the value; otherwise, the mix proportions should be either rejected or modified as per the advice of the concrete technologist.

6.2.4 Reinforcing Materials

Reinforcing materials are an important component of composites, and they are introduced into the materials with a purpose to improve their mechanical properties, especially post-cracking/yield response. It can be done by pasting them externally over the surface of the main member, or it can be embedded into the material. For instance, the reinforcement in the form of steel bars (called as rebars) or short, discrete fibers is commonly embedded in concrete to enhance its tensile strength or make it fit to support the load in bending. This type of the reinforcement is called as internal reinforcement. The other way round, it can be done by pasting some material which is strong in tension on the face of a main weaker member; e.g., steel plate can be pasted on the face of wooden beam (known as the fletched beam) to increase their flexural capacity. Similarly, the shear deficient concrete beams can be strengthened by pasting the steel plates on its web. This type of the reinforcement is known as external reinforcement. Like the concrete, soils can also be reinforced to improve their load carrying capacity and the response under the action of external loads; they normally carry the load by means of the direct contact and the internal friction. Geotextiles are used as reinforcing materials to make soil perform as per the requirement of design constraints. The reinforcing materials generally share a major chunk of the applied load and impart ductility to the composite. They mostly contribute by mobilizing the surface frictional forces in the form of the bond between them and the embedding material. For instance, it is the bond stresses between the concrete and the rebars that impart tensile resistance to the reinforced concrete. Similarly, it is between the geotextiles and the adjoining soil mass in case of reinforced soil that leads to an improved mechanical response of the composite material. Now-a-days, a variety of reinforcing materials are available, each possessing different strength and other characteristics. As the performance and capacity of the composites depend directly on the strength and other characteristics of reinforcing materials, it is highly desirable to select and provide reinforcing materials that have consistency in their characteristics. A deviation in the results more than the permitted values must lead to their rejection as a building material.

6.2.4.1 Structural Steel

In case of steel rebars, the chemical composition greatly controls their mechanical properties. A variation from the permissible values generally imparts brittleness to the rebars and affects significantly their yield strength and elongation characteristics. Table 6.12 gives the permitted chemical composition of steel rebars.

The rebars possessing this chemical composition generally exhibit the mechanical properties given in Table 6.13. These should not be allowed to deviate from the listed controlled values. In case of the deviations, the lot should be rejected and should not be used.

The tensile strength of steel rebars and their percentage elongation, the percentage total elongation at the maximum force and 0.2% proof stress is determined in accordance with the requirements of IS 1608, read in conjunction with IS 2062. The test results must meet the requirements given in Table 6.13. And, if the rebars have been found to comply with the listed provisions, they will be able to develop the required yield strength or ultimate strength only if they are bonded and anchored adequately with the adjoining concrete mass; the bond stress calculated under the prevailing load conditions must conform to the permitted values of the concrete used in the casting of the beams.

Table 6.12 Chemical composition of steel rebars, percent maximum values

Composition	Rebar grade								
	Fe 415	Fe 415D	Fe 415S	Fe 500	Fe 500D	Fe 500S	Fe 550	Fe 550D	Fe 600
Carbon	0.30	0.25	0.25	0.30	0.25	0.25	0.30	0.25	0.30
Sulfur	0.060	0.045	0.045	0.055	0.040	0.040	0.055	0.040	0.040
Phosphorus	0.060	0.045	0.045	0.055	0.040	0.040	0.050	0.040	0.040
Sulfur and phosphorus	0.110	0.085	0.085	0.105	0.075	0.075	0.100	0.075	0.075

Source https://archive.org/details/gov.in.is.1786.2008

Table 6.13 Physical properties of steel rebars, percent maximum values

Parameter	Rebar grade								
	Fe 415	Fe 415D	Fe 415S	Fe 500	Fe 500D	Fe 500S	Fe 550	Fe 550D	Fe 600
0.2% proof stress yield stress (N/mm^2) (Min.)	415	415	415	500	500	500	550	550	600
0.2% proof stress yield stress (N/mm^2) (Max.)	–	–	540	–	–	625	–	–	–
Elongation (%) (Min.)	14.5	18.0	20.0	12.0	16.0	18.0	10.0	14.5	10.0
Total elongation at maximum force (%) (Min.)	–	5	10	–	5	8	–	5	–
Ultimate strength and yield strength ratio (Min.)	1.10	1.12	1.25	1.08	1.10	1.25	1.06	1.08	1.06

Source https://archive.org/details/gov.in.is.1786.2008

The plain rebars posses smallest yield strength, and their plain surface texture compliment the requirement of smaller bond strength needed to develop the permitted value of the tensile force. However, in case of the high strength rebars (e.g., Fe500 and higher), ribbing of some geometric patterns on their surface becomes more important requirement which increases the bond strength manifold by mobilizing additional frictional forces. And, because of the high bond strength, they are able to mobilize a more tensile force as permitted by their higher yield strength. The size and areas of these ribbing patterns are fixed by various standards/codes to provide uniformity to the steel rebar producers, etc. The deformed high strength rebars will be considered fit for use in reinforced concrete members only if they are able to develop the bond strength in excess of 40 and 80% of that a comparable plain rebar (of same nominal diameter) would develop corresponding to the load observed at a measured slip of 0.025 mm and 0.25 mm, respectively. Moreover, the mean area of the ribs (in mm^2) per unit length (in mm) above the core of the rebar projected on a plane normal to its axis shall not be less than the following values, and the mean projected area of the transverse ribs alone should not be less than one-third of these values.

(a) 0.12φ for the rebar diameter (φ) < 10 mm
(b) 0.15φ for the rebar diameter range (φ), 10 mm < φ < 16 mm
(c) 0.17φ for the rebar diameter (φ) > 16 mm.

The reinforcing bars being the tension element contribute by mobilizing tensile force equal to the product of its available tensile strength and the cross-sectional area. The right chemical composition will ensure compliance of various strength parameters given in Table 6.13. However, during their production, any lapse on any account may lead to variations in the rebar diameter, which, in turn, will directly control the magnitude of tension force they would develop in the member. To keep a check on this part of the possible lapse, the mass of the rebar should not be allowed to deviate more than the permissible values. The rebar cross-sectional area (A, mm^2) can be determined from the weight measurements (in kg) done on a 0.5 m long rebar. It can be computed from Eq. 6.1.

$$A = \frac{w}{0.00785L} \tag{6.1}$$

where $\boldsymbol{w}$ is mass (in kg) weighed to a precision of ±0.5% and $\boldsymbol{L}$ is length (in m) of the test specimens measured to a precision of ±0.5%. The difference between the mass of two consecutive specimens shall be less than three percent; otherwise, the mass of the specimens ($\boldsymbol{w}$) shall be determined by removing their ribs and the area so computed from Eq. 6.1 shall be increased by three percent. This mass value must conform to the value (in kg) determined from the expression: 0.00785 kg/mm^2 per meter. The deviation, if any, must conform to the values given in Table 6.14.

Another important parameter that the rebars must qualify to allow their use in the reinforced concrete construction is their bending and rebending capability. It can be tested by doubling the rebar specimen over the mandrel until its sides become parallel. The diameter of the mandrel depends upon the rebar grade and its

Table 6.14 Maximum permitted tolerance on the nominal rebar mass

Nominal size (mm)	Batch (%)	Individual sample (%)	Individual coil (%)
Up to and including 10	±7	−8	±8
Over 10 and up to including 16	±5	−6	±6
Over 16	±3	−4	±4

Source https://archive.org/details/gov.in.is.1786.2008

Table 6.15 Mandrel diameter for different grades (= constant being tabulated × rebar diameter)

Rebar diameter	Rebar grade						
	Fe 415	Fe 415D	Fe 415S	Fe 500	Fe 500D	Fe 500S	Fe 550
Up to and including 10 mm	5	4	–	5	4	6	7
Over 10 mm	7	6	–	7	6	7	8
Up to and including 20 mm	3	2	4	3	5	4	5
Over 20 mm	4	3	5	4	6	3	6

Source https://archive.org/details/gov.in.is.1786.2008

diameter; it can be selected from the data sheet given in Table 6.15. The specimen is considered satisfactory if there is no rupture or cracks visible to a person of the normal or corrected vision on the bent portion of the rebar test specimen.

In the rebend test, the rebar specimen is bent to an included angle of 135° using a mandrel having an appropriate diameter (see, Table 6.15). The bent piece is kept in boiling water for 30 min, and then it is allowed to cool in air. The test piece is then bent back to have an included angle of 157°. It is considered to have passed the test if there are no ruptures or cracks visible to a person of the normal or corrected vision.

The test specimen prepared from the rolled steel sections if tested under the standard tension loading conditions (IS 1608) must meet the strength parameters given in Table 6.16. Test pieces should be cut crosswise from plates and strips and lengthwise from the sections, flats and bars. The geometric dimensions of the rolled section must conform to their standard sizes; the details are available in different codes for various sections, such as the beam, the column, the channel and the angle sections (IS 808); the tee bars (IS 1173); the plates, strips and flats (IS 1730); the round and square bars (IS 1732); the channel sections (IS 3954).

The steel test specimen is deemed to comply with the listed standards, if the average value of three test specimens meets the requirements given in Table 6.16, provided no individual value shall come less than 70% of the specified value during the testing. If the average value of the three Charpy impact tests fails to comply by an amount not exceeding 15% of the specified minimum average value, three additional test pieces from the same sample shall be tested and a new value of the

Table 6.16 Mechanical properties of different steel grades

Steel grade	Tensile strength (MPa)	Yield strength (MPa) for a test specimen thickness (*t*) of			Percent elongation, Min	Internal bend diameter, Min		Charpy impact test
		<20 mm	20–40 mm	>40 mm		<=25 mm	>25 mm	
E 250	410	250	240	230	23	2t	3*t*	27
E 275	430	275	265	255	22	2t	3*t*	27
E 300	440	300	290	280	22	2t	–	27
E 350	490	350	330	320	22	2t	–	27
E 410	540	410	390	380	20	2t	–	25
E 450	570	450	430	420	20	2.5t	–	20
E 550	650	550	530	520	12	3t	–	15
E 600	730	600	580	570	12	3.5t	–	15
E 650	780	650	630	620	12	4t	–	15

Source https://archive.org/details/gov.in.is (respective codes, such as IS808, IS1173, IS1730, IS1732, IS3954)

average should be calculated by adding the results to those obtained previously. If the new average value of the test result comply with the specified requirement, the material represented shall be deemed to comply with the listed provision of the applicable standard.

6.2.4.2 Fibers

These are the wires cut into short lengths and are mixed with aggregates and cement to produce concrete and reinforce it internally. The wires can be of steel or natural or any other polymeric materials. Generally, the steel fibers are used in the concrete to modify their mechanical properties, especially the residual tensile strength and the post-cracking behavior,, whereas the fibers manufactured from the polymeric and/or natural materials, such as nylon and jute, arrests the early-age microcracking caused by the shrinkage. Steel fibers are available in various shapes and forms. The most prominent among them are hooked-end fibers, crimped fibers and wavy-shaped fibers. The two major parameters that influence the mechanical characteristics of the steel fiber-reinforced concrete are the fiber aspect ratio and its volume fraction used in the mix, in addition to their shape. The concrete mix is generally designed to achieve certain strength levels and the performance characteristics, and accordingly, these two fiber parameters are varied at the design stage to get the desired performance from the concrete in the hardened state as well as in the fresh state.

Fibers are generally prepared from the cold drawn wire, cut sheet, melt extract or milled from blocks and cut into suitable shapes; straight or deformed. The steel fibers are then bundled together in the form of a number of fibers glued together using a water soluble glue, or wrapped as 'pucks' or supplied in the form of a belt to facilitate easy dosing and mixing. This helps to achieve consistency during the mixing operations. The anchorage of the steel fibers into the adjoining concrete mass in a structural member is a sole source of the improved mechanical properties exhibited by the concrete containing fibers. Every attempt, therefore, should be made to ensure consistency in the mixing and during the concrete placement operations as per the specifications agreed in the contract document. This is generally specified by means of either, (1) fiber type and content; expressed in terms of the fiber-aspect ratio and volume fraction; (2) by the member's performance level, expressed either in terms of the residual tensile strength/fiber-index, the permitted crack-width, the post-crack behavior at the ultimate and the serviceability limit states, early-age shrinkage, and consistency, etc. Out of these two approaches, the simplest and most common method for a designer is to define the fiber type and the fiber content per m^3 (specified as a minimum quantity) that should be included in the concrete to attain requisite strength properties. In this case, the responsibility of a concrete producer is limited only to adding the cement and the aggregates as per the mix proportions and mixing the right type, and the specified quantity of fibers in the concrete mix and then, ensuring that the fibers are mixed homogeneously therein, and finally, the other specified requirements with regard to the consistency

Table 6.17 Permitted tolerances for different fiber characteristics

Fiber characteristics		Deviation of individual value from declared value	Deviation of average value from declared value
Length	>30 mm	±10%	±5%
	≤ 30 mm	±10%	±1.5 mm
Diameter	>0.30 mm	±10%	±5%
	≤ 0.30 mm	±10%	±0.015 mm
Aspect ratio		±10%	±7.5%
Tensile strength		±15%	±7.5%

Source Author

and the target compressive strength are achieved. The contractor/designer takes responsibility for the desired concrete performance resulting from the addition of the steel fibers in the concrete mix, such as workability, pumpability, and post-cracking flexural tensile strength; whereas, in the second method, the concrete producer is solely responsible for the design and the performance of the concrete, including decisions on the fiber type and content as per the terms of agreement, and the mix proportions required for a desirable level of the concrete mix consistency, pumpability, strength properties, etc. Nevertheless, if the specifications are specified in this way, the test method and conformity procedures must be mutually agreed at the very start of the project with the concrete producer and it must be clearly mentioned in the contract documents to avoid problem at the later stage. In any case, the deviations if any should be kept minimum possible from the values adopted in the design mix of steel fiber-reinforced concrete and it should never exceed the limits given in Table 6.17.

The fibers meeting the requirements given in Table 6.17 should be tested in the laboratory conditions as per the provisions of RELIM [7] or CRN [2] to check the compliance of the strength parameters specified in the specifications, e.g., something like this: to attain a residual strength of 2.5 and 1.5 MPa at the CMOD of 0.5 mm and 3.5 mm, respectively corresponding to the fiber dosage of 30 kg/m^3 if steel fibers of some specified shape and aspect ratio are added to the concrete.

6.2.4.3 Geotextiles and Geogrid

These materials are generally used to reinforce the soils and improve their strength and settlement characteristics. The use of these materials makes the poor/problematic soil manageable, thereby enabling construction at the site/places that otherwise is unsuitable for construction. However, unlike the reinforcement in the concrete, these are generally used in the fabric form to be provided in single or multiple layers. They are also used to serve some other purposes like as a filtering media, to separate the different soil layers, or to protect some layers/slopes from the possible erosion, infiltrations, etc. Typically, they are made from polymeric materials like Polypropylene, Polyester, Polyamide (Nylon), Polyethylene, Polyvinylidene Chloride and are available in three basic forms: woven, needle punched and thermal bonded.

When using the Geotextiles as reinforcement, it is utmost important to ensure thorough the specifications that they must possess adequate short-term tensile strength, the creep limited tensile strength on the long-term basis, desired elongation at the design stress, the required puncture resistance (CBR or cone drop) and coefficients of friction. As geotextiles function by mobilizing the frictional forces between the soil particles and their own surface, it is important that their surface characteristics do not degrade with time; otherwise, they may start malfunctioning over a period of time and fails to serve the intended purpose. Moreover, if the due precautions are not taken during their storage and installation, they are liable to be rejected as it may result in the change of their surface characteristics. Geotextiles which are not ultraviolet stabilized should not be kept uncovered for more than 7 days after their installation at the sites. And, those which are stabilized for the ultraviolet radiations; it cannot be more than 14 days after the installation. The geotextiles rolls should remain covered while in storage to minimize any potential damages before their installation, possibly arising from the following factors:

- Damage because of tearing, excessive mud, wet cement, epoxy, etc
- Exposure to ultraviolet radiation, including direct sunlight
- Exposure to precipitation and chemicals that are strong acids or strong bases
- Temperatures beyond a range of −30–60 °C.

The project specifications should be framed by keeping these points in mind; any violation or non-compliance should simply lead to the rejection of that lot of the materials coming to the notice.

6.2.4.4 Carbon Composites

These are a family of composite materials which consist of a carbon (or graphite) matrix reinforced with carbon (or graphite) fibers that find applications as a good structural material due to its high strength, versatility and toughness. For instance, carbon–carbon composites with unidirectional reinforcement possess a tensile strength of 700 MPa and unlike the other available materials; it is able to maintain the strength even up to temperatures of 2000 °C. If by some means, the pyrolysis of the material is prevented, it can be used to sustain the load even at the higher temperatures. Unlike many other structural materials like the metals, the carbon composites have very high strength-to-weight ratio, the high heat resistance, low thermal expansion and excellent resistance to corrosion and radiation. These weigh only one-fifth of the metal weight. These are available as simple unidirectional fiber-reinforced structures to complex woven three-dimensional structures. As it is a special material, its acceptance criterion should be framed keeping in mind the requirements of the project and other design parameters.

6.2.5 Concrete

Unlike the carbon composites, the concrete is the most widely used structural material in both routine and very specific civil engineering applications. It can be produced at the site as well as in the highly mechanized plants to meet the desired mechanical and other material properties. Mixing and use of some other materials, such as admixtures, fibers of different types and shapes, rebars, and porous aggregates at the time of concrete production/concreting operations yields a variety of the mechanical properties and very distinct other material characteristics. The concrete so produced and placed is acceptable only if it meets some important material parameters; as otherwise, it may not be able to provide the requisite strength and meet the targeted other equally important properties in its fresh and hardened state. These are described below for different types of concrete.

6.2.5.1 Plain Concrete

Concrete without having any additional materials (like fibers, reinforcing steel), except the basic constituents of cement, water, fine and coarse aggregates is called as a plain concrete. To produce such concretes, the mix proportions should be so selected as to ensure the desirable workability of the fresh concrete and when concrete is hardened, it should posses the required strength, durability and surface finish. It can be designed to have any value of compressive strength; it all depends upon the type of applications or locations where the concrete is to be used. For instance, concrete of M25 grade or higher is recommended in the locations where it is expected that the concrete may come in contact with some aggressive chemicals during its life span; whereas M15, M20 grades can be used for the routine indoor protective environmental conditions.

Since the concrete is a heterogeneous mix of different materials, their individual chemical and physical characteristics will have a pronounced influence on the fresh and the hardened strength properties as well as the durability characteristics. A careful selection of suitable type and size is a prerequisite to get the desirable material response during its entire life span. The presence of chloride, sulfates and other similar chemicals in concrete beyond certain limits often leads to deterioration of concrete mass in the form of spalling, delamination, microcracking, etc.; these harmful salts may find their way into the concrete either by means of diffusion from the environment or from the basis constituents such as cement, aggregates, water and admixtures. Therefore, in order to prevent or minimize the chances of deterioration of concrete from the harmful chemical salts, their concentration levels should be limited to their permissible values.

The total amount of acid soluble chloride content in the concrete at the time of placing shall be limited to 0.6 (expressed as kg/m^3 of concrete) in the situations where some metal component is required to be embedded in the concrete mass; whereas it is limited at 3.0 for the concrete without any such embedded metal. The higher the chloride content, or if it is subsequently exposed to warm–moist conditions the greater the risk of corrosion of embedded metals will be present.

Sulfates are present in most of the cements; these are also found in some varieties of the aggregates. Excessive amounts of water-soluble sulfate can cause expansion and disruption of concrete surface. To prevent this, therefore the total water-soluble sulfate content from various constituents of the mix of the concrete mix (expressed as SO_3) should not exceed four percent by mass of the cement in the mix.

Similarly, certain aggregate types are very susceptible to attack by alkalis (such as Na_2O and K_2O) originating from cement or some other sources in the concrete and these very often leads to expansive reaction resulting cracking and disruption of the concrete. The extent of any such damage depends upon the moisture level within the concrete, the concentration of alkalis and the quantity of the aggregates containing an alkali reactive constituent. The presence of all these factors produces significant expansion within the concrete mass and that leads to the consequential internal microcracking and, possibly, delamination in the concrete member unless certain precautions (listed below) are taken to minimize the damages arising from all such chemical reactions:

- Use of non-reactive aggregate from some other alternate sources
- Use of low alkali cements having total alkali content not more than 0.6% (as Na_2O equivalent)
- Limiting the cement content in the concrete mix and thereby limiting total alkali content in the concrete mix
- Use of admixtures as a partial replacement of ordinary Portland cement in the concrete, such as fly ash or granulated blast furnace slag, provided their minimum dosage should be at least 20% and 50%, respectively
- Use of other measures to reduce the degree of saturation of the concrete during the service life, such as use of impermeable membranes.

In order to avoid the possibility of early thermal cracking or the cracking caused by drying shrinkage or any damage caused by the alkali silica reactions in concrete, the cement content in the concrete should not be allowed to exceed a value of 450 kg/m^3. Depending upon the environmental exposure conditions where the concrete is to be used during its intended life span, the minimum cement content should be ensured as per the lower limits reproduced in Table 6.18.

By selecting the basic constituent materials that meet these listed requirements, the concrete mix can be proportioned and designed to achieve any desired compression strength and other field constraints, such as workability and pumpability. Indian standard IS 10262 has detailed provisions to do this activity; once that is complete, the next task is to mix and produce the concrete. It must be checked and verified from the available documents that the dosages of retarders, plasticizers and superplasticizers if specified in the mix report should not exceed 0.5, 1.0 and 2.0%, respectively, by weight of cementitious materials.

Table 6.18 Minimum cement content in plain concrete

Exposure	Cement content (kg/ m^3)	Free water–cement ratio (Max.)	Min. concrete grade
Mild	220	0.60	–
Moderate	240	0.60	M15
Severe	250	0.50	M20
Very severe	260	0.45	M20
Extreme	280	0.40	M25

Source https://archive.org/details/gov.in.is.456.2000

Concrete should be mixed in a mechanical mixer complying with IS 1791 and IS 12119. This mixing should be carried out for a minimum period of two minutes or till the time it attains a uniform color, consistency and distribution of the materials. The concrete so produced must possess the required workability; it can be checked from the field report and it must be as per the project specifications or can be determined by performing the required test at the site, such as slump value or vee-bee time. Table 6.19 lists the suggested range of the concrete workability for the various routine applications. In situations where the slump of less than 25 mm is required, the workability should be determined by means of the compacting factor; a value of 0.75–0.80 should be achievable in all such cases.

Samples from the fresh concrete shall be taken as per IS 1199 and the cubes should be made, cured and tested at 28 days in accordance with the provisions of IS 516 to see the compliance of the mandatory strength requirements enshrined in IS 456. The concrete sample is deemed to comply with the strength requirements when both of the following conditions are met:

- The mean strength determined from any group of four consecutive test results having a standard deviation (s) compiles with the value computed from (f_{ck} + 0.825 s) or ($f_{ck} + \alpha$).
- Any individual test results complies with ($f_{ck} - \alpha$), where (α) is 3 for M15 concrete and is 4 for all other concrete grades above M15.

And, the standard concrete specimen must meet the following flexural strength requirements:

Table 6.19 Concrete slump values for different pouring conditions

Placing conditions	Degree of workability	Slump value, (mm)
Mass concrete; lightly reinforced sections in slabs, beams, walls, columns; floors; hand placed pavements; canal lining	Low	25–75
Heavily reinforced sections in slabs, beams, walls, columns, strip footings; slip-form work; pumped concrete	Medium	75–100
Trench fill, in-situ piling	High	100–150

Source https://archive.org/details/gov.in.is.456.2000

- The mean strength determined from any group of four consecutive test results exceeds the specified characteristic strength by at least 0.3 MPa.
- The strength determined from any test result is not less than the specified characteristic strength less 0.3 MPa.

The concrete compression strength as required in the above stated acceptance conditions can be determined by testing the standard concrete cubes of 150 mm size in a suitable loading machine at the age of 28 days. It must be ensured that the machine should be duly calibrated and capable of producing error free results. The permissible error should not exceed ±2% of the maximum load. The test specimens stored in water can be tested immediately on removal from the water and while they are still in the wet condition. However, the surface water and grit if any, should be wiped off the specimens. In case, the specimens are received dry, these should be kept in water for 24 h before they are taken for testing. The dimensions of the specimens to the nearest 0.2 mm and their weight shall be noted before testing. After that, the test specimen is placed between the jaws of the loading machine; it must be ensured that the axis of the specimen should be exactly along the loading axis to avoid any premature failure caused by the bending stresses in the specimen. The load shall be applied without shock, and it should be increased continuously at a specified rate of approximately 140 kg/cm^2/min until the specimen fails to hold any additional load. The maximum load that the specimen carried at its failure is recorded to compute its compressive strength; it can be determined by dividing the recorded value of the maximum compressive load by the actual measured cross-sectional area of the test specimen. In case, the strength values of the three cubes differ by more than 15% of their average strength, a new set of cubes should be tested to see the compliance.

Similarly, the flexure strength of the concrete can be determined using the standard prism of size 150 × 150 × 700 mm if the maximum sizes of coarse aggregates do not exceed 38 mm. The smaller prism size of dimensions 100 × 100 × 500 mm will be suitable to prepare the specimens in case the maximum coarse size up to 19 mm is used in the concrete mix. The hardened concrete specimen after 28 days of curing should be tested under four-point loading conditions in the wet condition with their surface wiped off any excess water; it should be made dry before testing them. The load is applied without shock and increased continuously at a specified rate such that the extreme fiber stress on the face of the specimen increases at approximately 7 kg/cm^2/min; that is, at a loading rate of 400 kg/min for the 150 mm-sized specimen and at a loading rate of 180 kg/min for the 100 mm size specimens. The specimens are loaded till they fail to carry any additional load; the maximum load carried by them at the failure is recorded to compute their flexural strength and is expressed in the unit of MPa.

6.2.5.2 Reinforced Concrete

Plain concrete containing steel reinforcing bars (called as rebars) in the form of longitudinal and the transverse steel is known as reinforced concrete. The rebars are added mainly in the member tensile zone to prevent the sudden loss of concrete strength after reaching certain limiting strain levels; these rebars in concrete sections impart them the required strength and ductility to withstand the imposed loads. All rebars must be free from loose mill scales; loose rust; coats of paints, oil, mud or any other materials which may destroy or reduce the bonding between the concrete and the embedded steel once the member is hardened to perform its intended functions. In case, if any such type of the layers (of any foreign material) is present over the rebar surface, sandblasting or other similar treatment may be carried out to clean their surface of any such superficial layers before the concreting operations begin.

The rebars should be cut and bent in accordance with the procedure prescribed in IS 2502, and subsequently, these should be placed and maintained in the position as is shown in the good-for-construction (GFC) drawings by providing proper cover blocks, spacers, supporting bars, etc. Normally, the spacing of spacers or chairs should not be kept at a spacing of more than 1 m to avoid any possibility of the change in the specified cover/spacing in-between the rebars during the concreting operations. The actual concrete cover to the rebars should not deviate from the specified nominal cover in the structural drawings by +10 mm. On-site decisions to weld the rebars, if required on any account, should be properly documented with all permissions of the design engineer and the project engineer-in-charge. In general, welding of rebars or use of mechanical connections (coupler) is permitted to join different rebars if found short in length than that required in the members, but in all such cases, the test results should be a part of the site record to prove that the joints are having the strength at least of the full strength of the rebars being connected. Welding of rebars having a diameter of 25 mm or higher is permitted and if the site conditions demand welding of two different rebars in the lap region, it should be done in accordance with the recommendations of IS 2751 and IS 9417.

The concrete must meet the basic specifications needed to produce a good quality plain concrete, as are prescribed in the previous section, plus some additional constraints on the presence of various harmful chemicals. The presence of certain chemicals in excess of the permitted values usually triggers corrosion of the embedded rebars in the concrete. Chlorides among them are the most important chemical that should not be allowed in the concrete mix in excess of its permitted range; the total amount of the acid soluble chloride content in the concrete at the time of its placing operations shall be as limited to 0.3 only (expressed as kg/m^3 of concrete). In addition to this, the minimum cement content and the maximum value of the water–cement ratio in the concrete mix should be restricted to the values listed in Table 6.20.

The minimum cement content in concrete as given in Tables 6.18 and 6.20 is for the concrete containing 20 mm nominal-sized coarse aggregates. If small-sized coarse aggregates (nominal size of 10 mm or 12.5 mm) are being used, then the

Table 6.20 Minimum cement content in reinforced concrete

Exposure	Cement content (kg/m^3)	Free water–cement ratio (Max.)	Min. concrete grade
Mild	300	0.55	M20
Moderate	300	0.50	M25
Severe	320	0.45	M30
Very severe	340	0.45	M35
Extreme	360	0.40	M45

Source https://archive.org/details/gov.in.is.456.2000

specified minimum cement content in the mix should be increased by 40 kg/m^3 and for the large-sized aggregates (over 40 mm nominal size), the specified cement content may be reduced by 30 kg/m^3. These points are the bare minimum that should be part of the material specification. Additional parameters/points must be added after due deliberations with the design team that leads to better quality and performance during the intended project service life.

6.2.5.3 Fiber-Reinforced Concrete

It is another type of reinforced concrete, which contains short fibers randomly mixed at the time of its production. The use of conventional reinforcing bars sometimes also becomes necessary to meet the required strength and ductility as well as the concrete workability. The fiber parameters, namely the fiber aspect ratio, volume fraction and its shape used to produce the concrete play a pivotal role in controlling the response of fiber-reinforced concrete in the hardened state as well as the fresh state. It is therefore utmost important to ensure that only the designed dosage of the fibers of a specified shape and aspect ratio should be used and mixed to produce a uniform and homogeneous concrete mix. The presence of fibers in the mix reduces its workability and makes the placement operations little bit difficult in comparison with the normal concrete with similar proportions so proper adjustments need to be made in the mix proportions to get the desired workability.

The concrete basic constituent materials should be of desirable characteristics as are outlined for the plain and the reinforced concrete in the previous sections. The compliance becomes necessary to avoid the possibility of any durability-related problems caused by the harmful chemicals finding their way into the concrete.

It must be cross-verified from the design records that the fiber-reinforced concrete being used for any structural applications should have been proportioned to exhibit a strain hardening response. In case of the strain-softening response under some exceptions, the characteristic value of the residual tensile strength (f_{R3}) should be kept at least 20% of the strength exhibited by the concrete at the service load. Either way, the conditions specified in Eqs. 6.2 and 6.3 (both) must be satisfied by the concrete containing steel fibers if the purpose is to replace the conventional rebars with the steel fibers:

$$\frac{f_{R1}}{f_L} \geq 0.4 \quad (6.2)$$

$$\frac{f_{R3}}{f_{R1}} \geq 0.5 \quad (6.3)$$

where f_L is the characteristic value of limit of proportionality (LOP) and f_{R1} and f_{R3} are the characteristic residual flexural strength of steel fiber-reinforced concrete corresponding to a crack mouth opening displacement (CMOD) of 0.5 mm and 2.5 mm, respectively. This will help to prevent the brittle failure of the section and also to reduce the crack-widths that would develop therein as well as enable them to achieve the minimum strength requirements.

The material residual strength as required in Eqs. 6.2 and 6.3 should be computed as per the provisions given in contract documents. In the absence of any clear-cut instructions in this regard, it may be carried out as per the standard testing guidelines, such as RILEM TC 162 [7] and CNR-DT 204 [2], albeit a prior acceptance should be obtained from all parties under the contract.

RILEM TC 162 [8] recommends the use of a set of beam specimens of dimensions 150 × 150 × 600 mm (with an effective span of 500 mm) with a mid-span notch of 5 mm wide or less for this purpose. The unnotched depth of the beam specimen should be kept as 125 ± 1 mm. The dimensions of the specimen shall not vary by more than 2 mm on all sides from the prescribed ones. It is also recommended that the difference in the overall dimensions on opposite sides of the specimen shall not be greater than 3 mm. The device for measuring the specimen dimensions should have an accuracy of 0.1 mm. This beam size can be used for a concrete having the steel fibers of length up to 60 mm and made using aggregates up to the size of 32 mm. It is required that the concrete should be compacted by means of external vibration only; however, in the case of a self-compacting steel fiber concrete, the mold shall be filled in a single pours and leveled off without any compaction.

The concrete specimens shall be cured in the molds for 24 h after the casting, at (27 ± 2) °C, either under a polyethylene sheeting or at not less than 95% relative humidity. The specimens are then demolded between 24 and 48 h after the casting and cured for a further period until preparation for the test. Testing under a three-point loading conditions shall normally be performed at 28 days unless otherwise is specified at some other time period.

The steel rollers with a diameter of 30 ± 1 mm shall be used to support the beam at its ends and also, at the load point on its top face. These shall be capable of rotating freely around their axis. The device used for measuring the load and the mid-span deflection (and also, for crack mouth opening displacement (CMOD) in the test setup) must have a least-count of 0.1 kN and 0.01 mm, respectively. The notch CMOD measuring system should be installed along the longitudinal axis at the mid-width of the test specimen, so that the distance between the bottom face of the specimen and the axis of the measuring system is 5 mm or less.

The specimen should be loaded gradually to produce a mid-span deflection at a constant rate of 0.2 mm/min until the specified final deflection is reached (normally, it is taken as 3 mm). The value of the load and the beam mid-span deflection should be recorded continuously. It is important to note that if the crack forms outside the notch, the test has to be rejected and it should be performed on a fresh beam specimen. When the test is executed by means of the CMOD control, the machine shall be operated in such a manner that the CMOD should increase at a constant rate of 50 μm/min for CMOD from 0 to 0.1 mm, followed by a constant rate of 0.2 mm/min until the end of the test. The test results should be plotted in the form of a load-crack opening response of the specimens. Figure 6.1 depicts a typical response of the concrete specimen containing steel fibers.

The residual tensile strengths F_{Ri}, shown in Fig. 6.1, are the key-parameters characterizing the post-cracking behavior of the concrete reinforced using the steel fibers. The load F is a load corresponding to the limit of proportionality exhibited by the test specimen. It can be taken as a load corresponding to the 0.5 mm mid-span deflection (δ) or a CMOD value of 0.05 mm. It is important to note that the load-deflection response of the member is linear-elastic up to the load F, and it may or may not be accompanied by the crack formation; therefore either of these two values can be adopted as the limit of proportionality.

Similar to the value of F, the values of other residual tensile strength parameters, such as F_{R1} and F_{R4} can be noted corresponding to a specified value of the mid-span deflection or CMOD value; F_{R1} is a load value corresponding to the CMOD of 0.5 mm or a deflection value of 0.46 mm; whereas F_{R4} is measured at a CMOD of 3.5 mm or a deflection value of 3.00 mm. It is worth to note that the value of f_{R1} can be less or higher than F_L; it all depends upon the fiber parameters taken in the concrete mix design. Once these strength parameters are obtained from the test data, the material should be checked against the approved specifications of the fiber-reinforced concrete; and, an appropriate decision may be taken regarding its acceptance or rejection.

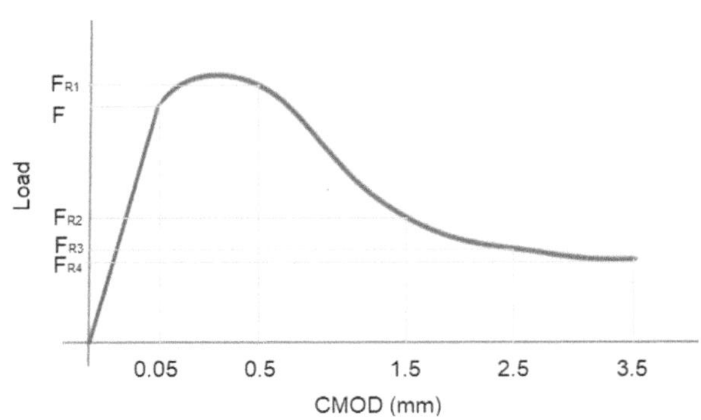

Fig. 6.1 A typical load-crack opening displacement response of steel fiber-reinforced concrete. *Source* Author

6.2.5.4 Self-Consolidating Concrete

It is a special type of concrete that possesses an ability to flow and fill the required space or formwork completely under its own weight. It does not need any external effort (such as vibrations) to achieve compaction after its placement in the molds. Unlike the plain concrete, it offers a rapid rate of concrete placement, with faster construction times and ease of flow around congested reinforcement in the members being poured. The fluidity and segregation resistance of the concrete ensures a high level of homogeneity, minimal concrete voids and uniform concrete strength, thereby providing a superior level of the surface finish and the material durability. The physical and chemical properties of the cement and other constituent materials of the concrete must be same as are required for the plain concrete and the reinforced concrete.

Self-consolidating concrete (SCC) possesses the ability to flow into and fill all spaces within the formwork, under its own weight; it is called as *filling-ability*. Its ability to flow through the tight openings such as spaces between the steel reinforcing bars without segregation and blocking is another characteristic, known as *passing-ability*. Whereas the ease with which it flows when unconfined by the formwork and/or the reinforcement defines its *flowability*. It also possesses an excellent ability to remain homogeneous in composition while in its fresh state, known as *segregation resistance*. While evaluating the SCC, it is therefore very important to ensure that the concrete to be used as self-consolidating must possess these four characteristics. Tests should be conducted to verify that the concrete is qualifying on these four parameters. The concrete mixture proportions are very critical to achieve these characteristics and require an optimized combination of coarse and fine aggregates, cement along with a carefully selected dose of chemical and other mineral admixtures. Generally, the coarse aggregates are reduced in the mix to increase the cement paste content. This process leads to a concrete mix with a good deformability and the high resistance to segregation and bleeding. A typical mixture composition of SCC is given in Table 6.21. It can be used to decide the initial SCC mix proportions and conduct various trials in the laboratory conditions till it meet the conformity requirements listed in Table 6.22, and accordingly, the concrete mix can be classified based upon the mutually agreeable test parameter between the supplied and the end user. The test methods to verify the acceptance of SCC must be a part of the agreement/contract.

In case, the trial mix fails to yield the desired mix proportions, following points can be tried to find the concrete mix that may conform to the acceptance criterion:

- Using additional or different types of filler
- Changing the proportions of the sand or the coarse aggregate in the mix
- Using a viscosity modifying agent if not used earlier
- Changing the dosage of the superplasticizer and/or the viscosity modifying agent
- Trying to adjust the water/powder ratio by changing the dosage of admixture in the mix.

Table 6.21 Typical mix proportions of SCC

Material constituent	Range by weight (kg/m^3)	Range by volume (l/m^3)
Powder	375–600	–
Water	150–215	150–215
Fine aggregates	45–55% of total aggregate weight	45–55% of total aggregate weight
Coarse aggregates	750–1000	275–375
Paste	–	300–375
Water/powder ratio (by vol.)	–	0.85–1.10

Source Author

The concrete (with its unique mix proportions) which meets the listed conformity criteria (see, Table 6.22) only qualifies as self-consolidating concrete and it can be used to fill the molds, on its own, of any geometrical shape and with any amount of reinforcing bars in the members being poured without imparting vibrators, etc., to achieve the desired compaction.

Table 6.22 Conformity criteria for the properties of SCC

Classification	Class defining ranges	Acceptance criteria	The test method to be used
Slump-flow class, SF1	550–650 mm	≥ 520 mm and ≤ 700 mm; $T_{500} < 3.5$ s	Slump-flow test
Slump-flow class, SF2	660–750 mm	≥ 640 mm and ≤ 800 mm; $T_{500} = 3.5$–6 s	
Slump-flow class, SF3	760–850 mm	≥ 740 mm and ≤ 900 mm; $T_{500} > 6$ s	
Slump-flow class specified as a target value		±80 mm of the target value	
V-funnel class, VF1	<8 s	≤ 10 s	V-funnel test
V-funnel class, VF2	9–25 s	≥ 7 s and ≤ 27 s	
V-funnel specified as a target value		±3 s	
L-box class (H_2/H_1 ratio)	≥ 0.8	≥ 0.75	L-box test with 2 or 3 bars
L-box specified as a target value		Not more than 0.05 below the target value	
Sieve segregation resistance class, SR1	$\leq 20\%$	$\leq 23\%$	Segregation resistance test
Sieve segregation resistance class, SR2	$\leq 15\%$	$\leq 18\%$	

Source Author

6.2.5.5 Lightweight Concrete

The concrete prepared to have smaller unit weight in comparison with the plain concrete is called lightweight concrete. Unlike the normal concrete, it weighs only 14–18 kN/m^3 caused by the uses of coarse aggregates mainly composing of the lightweight cellular and granular inorganic material, prepared either by expanding, calcining, or sintering products, such as blast furnace slag, clay, fly ash, shale, slate; or by processing some natural materials, such as pumice and tuff [6]. It must be ensured that the coarse aggregates selected for producing a lightweight concrete should contain clay lumps not more than two percent (by dry weight), nor these undergo a loss on ignition by an amount exceeding five percent. The chemical composition of various aggregates to be used in the production of a lightweight concrete should be similar to what is specified for the plain and the reinforced concrete in the previous sections. If the presence of harmful chemicals is found to be more than their permissible values, all such aggregates should be avoided to prevent any possible damage to the hardened concrete member.

Because of the use of relatively porous coarse aggregates in the production of lightweight concrete, its density unlike the plain concrete will not be a constant parameter. A new term called as equilibrium density is used instead. It is defined as the density reached by structural lightweight concrete after exposure to relative humidity of 50 ± 5% and a temperature of 23 ± 1.7 °C for a period of time sufficient to reach a density that has an insignificant change over the time (i.e., attain a near constant value). The specified equilibrium density can be converted into the maximum allowable fresh density by adding the density loss caused by the drying, which mainly depends upon the mixture proportions, air contents, water demand, particle density, and moisture content of the lightweight aggregate. The equilibrium density of a typical lightweight concrete mix is approximately 0.5–1.3 kN/m^3 less than its fresh density. As long as there is no appreciable change in the density value (i.e., not more than 0.65 kN/m^3 variation observed from the established fresh concrete density), there is reasonable certainty that the correct mix proportions are being batched. Otherwise, the mix proportion needs to be checked and corrective action is immediately required to change either the air content, or the moisture or the aggregate density to achieve the desired density. It should be regularly monitored to ensure that the concrete fresh density of any two consecutive batches reaching the site should not vary more than 0.5 kN/m^3 from each other, nor does it vary more than ten percent from the value considered in the design calculations.

The lightweight concrete can be used as structural materials only if it meets the one or more of the compressive strength requirements listed in Table 6.23 without

Table 6.23 Conformity criteria for using lightweight concrete for structural applications

Compressive strength, (MPa)	25	20	15
Unit weight (kN/m^3)	18.00	17.50	16.95

Source Author

exceeding the corresponding value of the maximum unit weight. Intermediate values for the strength and the corresponding unit weight values can be interpolated from the tabulated values.

Moreover, the total drying shrinkage exhibited by any such concrete mix should not exceed 0.10% and there should not be observable popouts/blowouts on the surface of the cast concrete surface caused by the release of entrapped air in the porous microstructure of coarse aggregates [1]. If such defects are appearing, there is a need to change the mix proportions. All these points must form the part of the material specification.

6.2.5.6 Permeable Concrete

The concrete made with a special purpose to pass water through itself is called as permeable concrete. It possesses a coefficient of permeability significantly higher than that of a normal or plain concrete, which enable water to drain to the underlying soils faster, thereby reducing the runoff from a site and allowing groundwater recharge. This type of the concrete is also called as no-fines concrete because of the near absence of fine aggregates in the mix proportion. Only coarse aggregates with little or no-fine aggregates are carefully mixed with the cementitious materials and water to get around 15–25% voids in the hardened concrete mass. The paste in the mix coats and binds the aggregates together to create a system of highly permeable, interconnected voids that allow a rapid drainage of the water in the concrete. Because of its high void ratio, the permeable concrete allows as much as 40–200 l/m^2/min of water to pass through the body of the concrete. The addition of a small quantity of the fine aggregates, typically 5–7%, is usually recommended to get some resistance against the possible degradation caused by freeze–thaw cycles. Moreover, depending upon the mix proportions adopted to produce the concrete, it possesses fresh density in the range of 1600–2000 kg/m^3. Table 6.24 shows a typical set of material consumption to decide the trial mix proportions of permeable concrete.

Unlike the normal concrete, the concrete density and its permeability expressed on the basis of the void content forms the basis of acceptance or rejection of the permeable concrete in any construction project. The concrete supplier is asked to

Table 6.24 Typical permeable concrete material proportions

Mix constituent	Common range to achieve the permeability characteristics
Coarse aggregate (kg/m^3)	1175–1475
Cementitious materials (kg/m^3)	265–400
Water-to-cementitious ratio	0.27–0.34
Aggregate-to-cementitious ratio (by mass)	4:1

Source Author

submit the most likely density of the proposed concrete mixture that will achieve the specified range of the target void content. This will accordingly fix the density of the concrete mix. For the acceptance purposes, the density of the fresh concrete being delivered at the site should be measured in accordance with the prescribed standard (usually, ASTM C1688); it should be measured at least once per day, or when some change is noticed in different lots of the concrete arriving at the site. The supplier should be alerted when a variation of more than ±80 kg/m^3 is observed in the test results of different lots as it may result in the change in the permeability characteristics of the concrete.

The actual void content in the concrete can be checked by conducting on-site test; wherein the permeable concrete is laid on the prepared subgrade surface and after 7 days, its ability to pass the required quantity of water through the designed thickness of the concrete panel is checked. If it is able to pass the required quantity of water in the stipulated time, then cores are taken from the test panel to determine the corresponding density of the concrete used to achieve the desired void content; the density is usually determined in accordance with ASTM C1754 or any other mutually accepted test method among the stakeholders. The average value of the density measured from the cores sets the basis for subsequent quality assurance and the acceptance testing during the casting operations. The concrete is fit to meet the qualities of permeable concrete or is acceptable for use if their density is found to be within ±5% of the average value obtained from the test panel.

As the strength of any material resides in its solid part, the presence of voids in concrete will reduce its compressive strength. Usually, the permeability of concrete increases with an increase in the void content/porosity; the permeable concrete is therefore expected to possess a lower compressive strength in the comparison with the normal concrete. Normally, it varies from 5 to 15 MPa; however, it is not uncommon to develop permeable concrete possessing compressive strength as high as 25 MPa. This range of the compressive strength indicates that it can be used as structural material as well.

The acceptance criterion should be therefore formulated by developing a reference point of the required concrete permeability, the density and the resultant compressive strength on the basis of the test results obtained from the cores taken from the test panel, as discussed in the previous paragraph. Trials can be made to achieve the required strength by changing the mix proportions and finding the resultant density and permeability. The concrete mix meeting the target can be adopted for the pouring operations. Alternatively, any mix achieving the target permeability and the density should be tested to determine the compressive strength on the basis of core taken from the sample and that value of the compressive and/or flexural strength can be adopted in the subsequent structural design of members/floor/pavement.

6.2.5.7 High-Performance Concrete

It is a very special type of the concrete usually produced to meet certain high end design constraints, such as requirements of high values of the strengths, toughness, elastic modulus, abrasion resistance, workability, durability and a long life in the severe environments along with an extremely low permeability, diffusion, shrinkage, creep and good self-consolidating characteristics. This type of concrete in all cases exceeds the properties and constructability of a normal concrete, thus making it a versatile material that can be used in all situations where the load levels are very high and the environmental conditions/expectations are very demanding. In order to produce such type of concrete, which performs exceptionally well on every performance parameter; engineers/producers need to be very careful while selecting the highest-quality ingredients and their optimized mixture designs and also during their batching, mixing, placement and compaction operations. A variety of the materials are employed to get high performance in comparison with the other types of concretes; a few such materials that are commonly used to achieve the design constraints of a high-performance concrete are listed in Table 6.25.

As this type of concrete is proportioned and produced for the specific project requirements, its performance in the fresh as well as hardened state should therefore be evaluated as per the identified performance parameters. It is always better to agree (among all stakeholders) and freeze these parameters along with their testing standards well before the starting the execution of the project. A typical set of such parameters is listed in Table 6.26 for the guidance of the user. It is not an exhaustive list; the parameters and their target values should be finalized as per the field/design constraints and the special requirements of the project if any. Whatever set of parameters are finalized; these will constitute the acceptance criterion. It is also

Table 6.25 A typical set of materials commonly used to produce high-performance concrete

Material	Purpose of the use of said material in the mix
Portland/blended cements, fly ash, slag, silica fume, calcined clay and shale, metakaolin	Cementing material/durability/high strength (both early and long term)
Optimally graded aggregate	Improve workability and reduce paste demand
Superplasticizers	Flowability
High-range water reducers	Reduce water-to-cement ratio, cement and water content
Retarders	Control setting
Accelerators	Accelerate setting
Corrosion inhibitors	Control steel corrosion
Shrinkage reducers	Reduce shrinkage
ASR inhibitors	Control alkali–silica reactivity
Polymer/latex modifiers	Durability

Source Author

Table 6.26 A typical set of materials commonly used to produce high-performance concrete

Performance parameter	Value that can be specified as target parameters	Testing standards	
		BIS guideline	Alternate testing guidelines
Compressive strength	Any value up to 140 MPa at 28 days, as per the design requirements	IS 516	ASTM C 39
Toughness	As per the design requirements		RELIM TC-162 CNR DT 204
High-early compressive strength	20–28 MPa at 3–12 h	IS 1199	ASTM C 39
High-early flexural strength	2–4 MPa at 3–12 h	IS 1199	ASTM C 78
High modulus of elasticity	More than 40 GPa	IS 1199	ASTM C 469
Residual tensile strength	As per the design requirements		RELIM TC-162 CNR DT 204
Density/unit weight	As per the specifications	IS 1199	ASTM C1688
Flowability	More than 600 mm	IS 1199	
Sieve segregation resistance	Less than 15%	IS 1199	
Abrasion resistance	0–1 mm depth of wear	IS 9284, IS 516(9)	ASTM C 944
Low permeability/ carbonation	500–2000 coulombs	IS 3085	ASTM C 1202
Chloride penetration	Less than 0.07% at 6 months	IS 516(12)	AASHTO T 259 & T 260
Resistance to chemical attack	No deterioration after 1 year solution in wet/dry environment	IS 14959	Concrete exposed to the saturated solution
Sulfate attack	0.1% expansion at 6 months for moderate sulfate exposures or 0.5% expansion at 6 months for severe sulfate exposure		ASTM C 1012
Resistance to freezing and thawing	Max. mass loss or expansion can also be specified		ASTM C 666
Resistance to deicer scaling	Max. mass loss or expansion can also be specified		ASTM C 672
Low creep	Less than normal concrete	IS 516(7)	ASTM C 512
Low shrinkage	Less than 400 millionth	IS 516(6)	ASTM C 157
Low absorption	2–5%	IS 1199	ASTM C 642

Source Author

equally important to freeze the relevant testing standards to see the attainment of the accepted criterion for all shortlisted design parameters; a list of related BIS standards are given in the bibliography section of the chapter.

6.2.6 Timber

Wood is a porous and fibrous structural tissue found in the stems and roots of trees. It is basically an organic polymeric material—a composite of the cellulose fibers that are strong in tension and are embedded in a matrix of the lignin present therein that imparts them compression resistance too. In a living tree, it performs a support function, enabling trees to grow large and stand up on their own. It also conveys water and nutrients between the leaves, other growing tissues and the roots. And, the wood processing into logs, beams and planks is known as timber. It finds many applications as a structural material well as serve many functional uses in building construction activities: wooden joinery, shuttering, scaffolding and furniture making, etc. Timber is classified into three groups, namely Grade-1, Grade-2 and Grade-3 based upon the presence of the permissible defects and cut sizes. It is worth to note that the timber belonging to the Grade-3 is not suitable for use as a structural material and if some material belongs to this grade is noticed in the procurement, it should not be accepted for any structural use. Moreover, the timber having loose grain, splits, heartwood rot and sap rot, crookedness and worm holes should be rejected IS 3629 [5].

In additional to this major requirement of the strength, the durability of the timber members must be specified/ensured through their different classes. It is classified into three major classes depending upon their life determined through a grave yard test (IS 401, 1982), namely Class-I (with an average life of 120 months or higher, it is timber of high durability); Class-II (with an average life between 60 and 120 months, it is timber of moderate durability); and Class-III (with an average life less than 60 months, it is timber of low durability). It is always advisable to go only for the timber exhibiting a moderate or high durability in order to conserve the natural resources and for a better economic utilization. However, it is always better to get the timber classes-II or III be treated to make it at par with the Class-I. Based upon the absorption of the conservator under a pressure of 1.05 MPa, out of these classes of the timbers only the timber belonging to the classes-I and II should be preferred, because only these two timber classes are treatable and can thus be preserved to stand for longer durations of their service life. It should also be ensured that the timber should not contain sapwood of more than 15%; otherwise, it must be thoroughly impregnated with wood preservatives to make suitable for structural applications.

The timber grades-1 and -2 are further grouped into certain sets depending upon their mechanical properties. These are groups-A, B and C with the first group possessing an elastic modulus and along the fiber tensile strength of more than 12,600 MPa and 18 MPa, respectively; while the group B has these values as

9800–12,600 MPa and 12 MPa, respectively. The group C has these values less than 9800 MPa and 8.5 MPa, respectively. Because the timber beams or any other structural members are generally designed on the basis of these two mechanical properties, it is therefore very essential that only the timber belonging to the required group should be a part of the procurement. Moreover, in order to get the desired strength properties, the timber must be thoroughly seasoned to limit the moisture content less than 18%.

Among the other factors that need due attention while accepting timber for structural purpose is the slope of the grains; it should be limited to 1 in 15. This is because of the fact that the timber offers maximum strength along the grains and it reduces if the slope of more than the permitted. In such cases, the design must be modified to account for the excessive slope into the calculations. As far as possible, it should be tried to have timber without knots and it must contain at least 12 rings per a strip of 50 mm. The knots if present, their size should be within the permitted limits; it should not normally exceed 15% of the member width.

After an appropriate timber is procured and is ready for cutting and installation, the variations in their geometric dimensions should be kept with the permissible tolerances. The length of the member must be with the tolerance of—0 and +10 mm, while its width or depth (if it is less than 100 mm) is kept within the range of—0 and +3 mm. For the member cross-sectional dimensions of more than 100 mm, the variations should not be more than—3 and +6 mm. The cut pieces if are found to have variations more than the permitted, these should not be allowed to be used for that particular member, and an attempt should be made to reuse these pieces by dressing them to cross-sectional dimensions of some other members smaller in size.

6.2.7 Glass

Glass is a non-crystalline material, often transparent and amorphous in nature, formed by a rapid cooling of the molten materials having high silica (SiO_2) content. It is a special type of the materials which at the atomic-scale shares characteristics of a supercooled liquid and at the macro-level exhibits mechanical properties of a solid. It forms naturally as well as can be produced artificially. For instance, glass forms naturally when the volcanic magma with a high silica content cools rapidly by coming in contact with water, e.g., Obsidian glass; on the other hand, it also forms when lightning strikes sand containing quartz through the process of vitrification. Artificially, it can be produced in mechanized plants and poured, extruded and molded into various forms ranging from flat sheets to other highly complex shapes. The finished products are usually brittle and fracture easily, unless these are tempered to modify their characteristics. Unlike many other materials, glass possesses special ability to refract, reflect and transmit light. The transparency in the glass results from the absence of the grain boundaries in the glass microstructure which otherwise diffusely scatter the light in case of the polycrystalline materials. It is also inert and resistant to chemical attack, thereby making it to the list of materials that can be used in the building where transparency and shield from the air movement is among the major requirements. It can also be used as a structural material in the laminated/toughened form. For instance, it can be used as a building cladding where it has to withstand the wind pressure along with other lateral loads that may act over the building facade. Like the metals, glass is very strong in tension and compression; nevertheless, its tensile strength can be further increased by means of tempering, etc. Table 6.27 lists few common properties of different forms of glasses.

As the glass sheets are produced in certain finite thickness, therefore the glass panels of some maximum dimensions only can be cut and used to cover the given opening size. The size may vary depending upon the safe stress value of the given glass type. As indicated in Table 6.27, the tempered glass needs a smaller thickness for any given opening size because of its relatively higher tensile strength in comparison with the plain or the laminated glass sheets. An indicative list is given in Table 6.28 that can be used to estimate the maximum panel size that can be cut for any given glass thickness.

Table 6.27 A typical set of mechanical properties for different glass types

Property	Typical value of different glass types		
	Plain/annealed	Tempered	Laminated
Density (g/cm^3)	2.42–2.52	2.42–2.52	2.42–2.52
Tensile strength (MPa)	40	120–200	30–40
Compressive strength (MPa)	1000	100	1000
Modulus of elasticity (GPa)	70	70	70
Coefficient of linear expansion (m/m K)	9×10^{-6}	9×10^{-6}	9×10^{-6}
Available thickness (mm)	2–19	3–19	4.38–12.38

Source Author

Table 6.28 Maximum possible glass sheet size area for any given value of the sheet thickness

Glass thickness (mm)	Maximum glass sheet size (m^2) in a framed panel consisting of		
	Plain/Annealed glass	Tempered glass	Laminated glass
3	0.1	1.0	–
4 (+0.38 for laminated sheet)	0.3	2.0	1.0
5 (+0.38 for laminated sheet)	0.5	3.0	2.0
6 (+0.38 for laminated sheet)	0.9	4.0	3.0
8 (+0.38 for laminated sheet)	1.8	6.0	5.0
10 (+0.38 for laminated sheet)	2.7	8.0	7.0
12 (+0.38 for laminated sheet)	4.5	10.0	9.0
15	6.3	–	–
19	8.5	–	–

Source Author

Table 6.29 Permitted thickness variations for different glass types

Glass type	Standard glass thickness (mm)	Permitted glass thickness (mm)	
		Maximum	Minimum
Plain/annealed and tempered glass	3	3.2	2.8
	4	4.2	3.8
	5	5.2	4.8
	6	6.2	5.8
	8	8.3	7.7
	10	10.3	9.7
	12	12.3	11.7
	15	15.5	14.5
	19	20.0	18.0
Laminated glass	5.38	5.81	4.95
	6.38	6.81	5.95
	8.38	8.81	7.95
	10.38	10.81	9.95
	12.38	12.81	11.95

Source Author

Once the thickness of the glass panel/sheet is decided or procured, the variations within the lot should not be more than the tolerances given in Table 6.29. In case, it is found to be more than the permitted values, the lot should be rejected; else the design should be revisited to see the possible changes in the panel size that can be made by adopting the glass thickness one size less than the available thickness for any given panel opening size and the applicable wind pressure acting thereon.

If the glass is required for its transparency and other optical characteristics, the specifications must be written to meet all those identified parameters. Non-conformance should simply lead to the rejection as the glass otherwise will not be able to achieve the intended functions and design goals.

6.2.8 Aluminum

Unlike the steel and most of other metals being used as structural materials, aluminum is the lightest among them and find enormous applications in the field of civil engineering, such as long-span roof systems in halls and auditoriums; structures having moving parts, such as sewage plant crane bridges and moving bridges; structural system for special purposes, for which maintenance operations are particularly difficult and limited, as in case of masts, lighting towers, antennas towers and so on. The smaller unit weight of this metal makes it possible to transport the fully prefabricated components directly to the sites and simplify the erection phases during the fabrication stage. It also helps to reduce the dead load being transmitted to the foundations during the service life of the built structure and help to reduce the cost associated with the labor and erection of structural systems. Another plus point of using aluminum as a structural material in the building and other built structures is its corrosion resistance properties that help to reduce the maintenance expenses and other costs associated with the maintenance and repair of different components.

It is worth to note that the pure aluminum metal is rarely used as a structural material due to its softness and low melting point, and the strength. It is also very ductile and possesses yield strength of around 10 MPa only, with a percent elongation of 50–70%. Whereas the different aluminum alloys have yield strengths ranging from 200 to 600 MPa, with the almost same unit weight as possessed by the pure aluminum metal. Excellent malleability and their low melting point make it easier to extrude sections with a large range of geometrical properties which simultaneously give the minimum weight and the highest possible structural efficiency [143]. However, because of their lower fatigue strength compared to the steel, aluminum and its alloys are never used in the conditions where considerable fluctuations in the stress/loading cycles (more than 10^7 stress cycles) is expected during the service life of a structural member. Aluminum alloys do not have this lower fatigue limit, and they continue to weaken with the continued stress cycles in contrast to the steel which exhibits a clear-cut lower limit, which is the cyclic stress amplitude below which no failures occur.

Aluminum alloys, moreover, have an elastic modulus of about 70 GPa, which is about one-third of the elastic modulus of most kinds of the steels and other steel alloys. Because of this small value of the modulus, the structural members being fabricated in this metal will exhibit higher deflections in the elastic range than a steel component of identical size and shape. But availability of different section shapes and sizes help to manage the problem to some extent by playing with the moment of inertia of the sections as it is the product of elastic modulus (E) of the material and moment of inertia (I) of the section made from such material that decides the deflections; the smallness of one parameter, e.g., **E**, can be compensated by using a higher value of the second multiplying factor (e.g., **I**) to get the desired value of the flexural rigidity (EI). The deflection being exhibited by the members is inversely proportional to its flexural rigidity for any given set of the loading and span.

There is another point that needs very high attention while designing some members using the aluminum as a structural material; it is its coefficient of the linear expansion and the creep coefficient which is almost double of the value that the steel have. A high value of coefficient of linear expansion of the aluminum causes the members to expand and contract relative to the dissimilar metals being used at the joints, eventually loosening of the connections. A similar problem exists with respect to the creep which leads to differential shortening of the members under a steady sustained load, again loosening the connection. Despite these major material limitations, aluminum alloys find enormous applications in different areas depending upon their characteristics. Table 6.30 lists a few such material grades along with their applications range and the available forms that can be of some use in selecting an appropriate grade of the aluminum.

Table 6.30 Different aluminum grades and their suggested application range

Aluminum grade	Characteristics	Available forms	Typical applications
19,000	Commercially pure aluminum, very ductile in annealed or extruded condition, excellent resistant to corrosion	Sheet, plate, extruded tube, wire, forgings	Architectural and builders, hardware, fasteners, paneling, molding welding wires equipments for food, chemical and brewing industries
19,500	High-purity aluminum, more resistant to corrosion than other grades	Sheet, plate, extruded tube, rolled rods, forgings	Marine fittings, pressed and anodized utility items, heat exchanger tubes, condenser tubes, gas and oil transmission pipeline
24,345	Strong alloy that is aged naturally at room temperature after solution treatment and has fair ductility in this condition	Sheet, plate, extruded tube, wire, forgings	Stressed parts in aircraft and other structures where high strength is of primary consideration, hydraulic tubes
31,000	Stronger than 19,000 have good weldability and corrosion resistance	Sheet, plate, extruded tube, rolled rods, forgings	General purposes alloy for moderate strength applications, pressure vessels, irrigation tubing, heat exchangers, utensils and pressure cookers, roofing sheets, pilfer proof and detonator caps, air-conditioning ducting fan blades and vehicle paneling
52,000	Ductile in soft condition, but work hardens to rapidly becoming extremely tough, has high resistance to corrosion in the marine atmosphere	Sheet, plate, extruded tube, wires, forgings	Paneling and structures, hydraulic tube appliances, refrigeration tubing condenser and heat exchanger tubes gas and oil transmission pipelines
63,400	Suitable for intricate extruded sections of medium strength, forms in W condition, highly resistant to corrosion	Extrusion, wire, tube rolled rod and forgings	Architectural uses, such as window/door frames, wall facings, partitions, and hand rails and other similar applications where surface finish is important and medium strength would suffice
64,430	A medium-strength alloy with good mechanical properties, corrosion resistance and weldability	Extrusion, wire, tube rolled rod and forgings	Structural applications of all kinds, such as road and rail transport vehicles, bridges, cranes, roof trusses and rivets. Cargo containers, milk containers, deep drawn containers and flooring

(continued)

Table 6.30 (continued)

Aluminum grade	Characteristics	Available forms	Typical applications
65,028 and 65,032	A medium-strength alloy with good mechanical properties, corrosion resistance and weldability	Extrusion, wire, tube rolled rod and forgings	Structural applications of all kinds, such road and rail transport vehicles, bridges, cranes, roof trusses, furniture, gas and oil transmission pipelines, condenser and heat exchanger tubes, hydraulic tubes
74,530	Medium-strength self-aging weldable alloy. It does not require heat treatment after hot working or welding. Excellent welding characteristics and good formability. Good corrosion resistance when compared with other high strength aluminum zinc alloys	Extrusion, plates, wire, tube rolled rod and forgings	Stressed structural applications requiring welding, such as bridges, checkered plates, dump-truck bodies, pressure vessels and rail coaches

Source https://archive.org/details/gov.in.is.738.1994

The material specifications therefore need to be very specific to the grade. Any variation will have a significant influence on the member's performance, and thus, it should simply lead to the rejection of the material lot coming to the notice of the supervisor.

6.2.9 Asphaltic Materials

These are the special type of organic material that posses the binding properties because of its inherent nature. Bitumen is the most common material from this family of the materials which is prepared by the refining of the crude petroleum. It is a sticky, black and semisolid form of the petroleum; it softens easily with the increase of temperature and, usually, remains in the semisolid form at the room temperature. This unique characteristic makes it very useful to keep the other material particles, such as aggregates and sand, if mixed with the asphaltic materials (bitumen), bound together at the normal working temperatures. Because of this unique nature, it is the most common and the oldest material being used in various construction activities, such as in road and industrial parking floor.

Based on the viscosity at 60 °C, the bitumen is graded into four groups, namely VG10, VG20, VG30 and VG40. While procuring some suitable grade of bitumen, the prevailing maximum air temperature in the area of the material is pivotal to the selection process. The prevailing temperature in the region should be 7 day average maximum air temperature taken for a period of not less than 5 years from the start of the design period of the project. If this condition is not satisfied, the paved surface may not perform as is required in the design; it will soften or become very hard/brittle, thereby failing to bind the mixed particles together. If some improper grade comes to the notice during the execution, the lot of the bitumen should straight away be rejected. The bitumen grade VG10 is the most suitable in the area if the temperature remains less than 30 °C; whereas the grade VG20 is the most appropriate for a temperature range of 30–38 °C in the area under considerations. On the other hand, the grade VG30 should be used if the temperature remains between 38 and 40 °C and VG40 is for the temperature of more than 40 °C. Moreover, the bitumen grade should remain homogenous in their composition and do not foam when heated to 175 °C and also, should be completely soluble in the Trichloroethylene. The other important parameters that the selected grade of the bitumen must satisfy are given in Table 6.31.

The bitumen is considered fit for the use, if its penetration value and the softening point determined from the equation: ($X \pm 0.6R$) is found to be as per the qualifying parameters given in Table 6.31, where X and R are the mean and the range of the test results, respectively. An appropriate grade of the bitumen when mixed with the graded particles such as sand, and aggregates produces a solid mass almost similar to the normal concrete to transfer the applied load to underlying media/strata without losing its integration; this mixture of the bitumen and aggregates is called as bitumen concrete. It is normally produced by heating the bitumen/asphalt binder to decrease its viscosity, and drying the aggregate to remove moisture from it prior to mixing. The mixing is usually carried out with the aggregates heated to about 150 °C and that the bitumen to a temperature of around 95 °C. An attempt is made to complete the paving and the compaction operations while the mix is sufficiently hot to achieve the required density. Otherwise, the material should be removed and reprocessed to be used again. But, it is always advisable to go for the fresh material prepared and laid exactly as per the identified specifications [3].

Table 6.31 Different qualifying parameters for different bitumen grades

Test parameter	Bitumen grade			
	VG10	VG20	VG30	VG40
Application range, based on 7 day Max. air temperature over a period of 5 years, °C	<30	30–38	38–40	>40
Softening point, °C, Min	40	45	47	50
Penetration at 25 °C, 100 g, 5 s, 0.1 mm, Min	80	60	45	35
Ductility at 25 °C, cm, Min	75	50	40	25
Absolute viscosity at 60 °C, Poises	800–1200	1600–2400	2400–3600	3200–4800
Kinematic viscosity at 135 °C, cSt, Min	250	300	350	400
Flash point (Cleveland open cup), °C, Min	220	220	220	220

Source https://archive.org/details/gov.in.is.73.2006

6.2.10 Soils

It is the earliest known natural material in use for various purposes. Almost everything man has constructed is over the soil/ground. Usually, it contains a large variety of particles ranging from a few microns to as large as 10 mm. Larger sizes are also present, but their proportions are generally smaller in the normal soil mass that occurs naturally in the top layer of crust. As described in the first chapter, the soils mainly mobilize its resistance to carry the applied load by means of the direct contact and the frictional forces that generates between the adjacent soil particles as they deform under the effect of the load. So naturally, a soil mass containing a large pool of different sized particles will be able to do so in the best way in comparison with the soil mass having particles of almost uniform size. The soil mass having particles of almost all sizes is called as well-graded soil, and in case, it contains particles of almost the same size or some of the sizes are missing is known as poorly graded and gap graded, respectively. The second possibility will be the case when the soil contains mostly extremely fine particles—small enough to generate cohesion to keep the flock of particles in the lumped form; it is classified as clayey soils. Usually, the clayey soils have particles of diameter less than 0.002 mm. When the soil has frictional characteristics as well as cohesive properties, it is called silty soils, which contain particles in the range of 0.002–0.05 mm.

Whenever the soils containing large coarser fraction are used as an engineering material, it is the prime requirement that it must be well-graded rather than a uniform or poorly or gap graded soil. If the available soil is not as per the criterion of a well-graded soil, then it must be blended with the soils, may be two or more, containing other particle sizes. It can be ensured by sieving the soil sample and based upon the results, calculating the coefficients of the uniformity (c_u) and curvature (c_c). For a sandy soil to be classified as well-graded, the sample must have a coefficient of uniformity of more than 6; while its coefficient of curvature must lie between one and three. Otherwise, it needs to be blended with other particles ranges to bring the values within the defined range else it is classified as poorly graded and should not be used as a structural material. Similarly for the gravelly soils (having particle size of more than 4.75 mm), the coefficient of uniformity of more than four and its coefficient of curvature must lie between one and three is okay to classify it as a well-graded soil mix. If the value of the coefficient c_c is found to be around 0.1 or less, it is indicative of poorly graded soils and if the value of both coefficients is computed as unity, it is a single-sized soil; the soil belonging to these two classes should never be used to transfer the loads. The values of these two coefficients can be computed from the following expressions by using the data from the sieve analysis.

$$c_u = \frac{D_{60}}{D_{10}} \quad \text{and} \quad c_c = \frac{(D_{30})^2}{D_{10} \times D_{60}}$$

where D_{60} is the soil particle diameter at 60% passing; D_{30} is the diameter at 30% passing, and D_{10} is the grain diameter at 10% passing; these three values can be picked up from the results of a sieve analysis of soil samples.

Almost all engineering properties in case of coarse-grained soils depend upon its gradation. Any soil found to be well-graded can be designed to meet any other requirements/design constraints of the project. The acceptance criterion will be a set of constraints that have been used in the design; e.g., it can be optimum moisture content (OMC) and the corresponding maximum dry density (MDD); whatever is specified in the specifications must be ensured to achieve on the site.

However, in case of silty or clayey soils, the presence of moisture in the soil defines their behavior. The very small particle size is capable of holding the water around their surfaces, which exerts significant influence on their response to the loading. Depending on its water content, a soil may appear in one of four states: solid, semisolid, plastic and liquid. In each state, the consistency and the response of a soil is different and consequently, so are their engineering properties. There exists a clear-cut boundary between each state where the soil undergoes a visible/observable change in its response to the mechanical disturbances. The Atterberg limits are very easy way to distinguish between different types of silts and clays; the water content at which the soils change from one state to the other defines the Atterberg limit. These are shrinkage limit, plastic limit and liquid limit of soils.

The shrinkage limit is the water content where further loss of moisture will not result in more volume reduction of a soil sample; whereas the plastic limit defines the moisture content where the soil can be made to a thread of about 3 mm when rolled over a flat, non-porous surface. On the other hand, the liquid limit denotes the water content at which the behavior of a clayey or silty soil changes from the plastic state to the liquid state. These limits are used in a number of ways, such as to the soil classification, to estimate the other soil properties such as its compressibility, permeability and strength. For instance, a difference between the liquid limit and the plastic limit of a soil, known as plasticity index, is commonly used to identify the suitability of soil for some engineering purpose. Soils with a high value tend to be clay, while with a lower value behaves more like a silty soil, and those with a zero value indicates sandy type soil with little or no silt or clay content. Soils with a plasticity index lower than 10% and a liquid limit lower than 20% behave as cohesionless soils. Normally, a soil having a plastic index of less than 7 indicates slightly plastic type soil; and within the range of 7–17 is a medium plastic and having a value of more than 17 is highly plastic in nature and it would behave like a clayey soil. Each soil type, whether sandy, silty, clayey or mixture of these exhibits entirely different responses and characteristics. As it is not practically feasible to replace the entire soil at the site of a project unless it is highly problematic or weak in nature, so while checking the acceptability of the given/available soil type, the specifications and the other parameters considered in the design should be checked and verified. If a large variation is observed between those taken into the calculations and that found in the soil being used for the construction, corrective actions become mandatory.

If the soil is to be borrowed to fill the low lying areas or for constructing some embankments, etc., the soil belonging to the non-plastic category should be preferred and an attempt should be made to make it well-graded by blending two or more types of different soils in order to achieve the desirable strength and other properties. For example, if we are constructing embankment without a clay core to pond water, the soil having a plasticity index in the range of 8 and 20% should be selected, and for achieving the best compaction results, it should be taken as close to 17% as for as possible; it can be done by making trials of mixing the different soils and that combination which gives the desirable value can be the one of the choices. The soil is usually compacted at its optimum moisture content (OMC) which needs minimum effort to achieve the maximum possible density; the optimum moisture content of soils is usually 2–3% less than the plastic limit. The soils exhibit maximum strength at its maximum dry density. On the other hand, for constructing the embankment of impermeable clay core, the ideal soil will be the one which possesses a plasticity index greater than 25% as the higher the clayey content more difficult will be for the water to seep through the soil mass, but at the same time, it may pose some other construction-related problems which can be handled by adopting the best available practices.

In addition to the strength, the soil compressibility is another property that must be paid due importance while selecting the type of soil; it must be a part of the specification. It denotes the degree to which a soil mass reduces in volume under the action of an imposed load. It is the lowest in coarse-grained soils where particles are in direct contact with each other and it increases as the proportion of small particles is increased in the soil mass and becomes highest in fine-grained soils having a major chunk of organic matter. For instance, gravels and sands are practically incompressible. If a moist mass of these two soil types is subject to some load, there is no significant change in their volume. Clays, on the other hand, are compressible in nature; if a moist mass of clayey soil is subjected to some loading, moisture and air present therein are expelled from it and the resultant volume changes in the soil are not immediately recoverable when the load is removed. Such type of soils needs engineering intervention to make them suitable for projects. Ground improvement is one such technique that is usually followed to improve the soil nature that fits the design constraints.

Conflict of Interest This chapter does not constitute a standard, specification, or regulation for which the applicable specifications, guidelines may be referred while freezing the construction-related documents.

References and Bibliography

References

1. ACI 213R Guide for structural lightweight aggregate concrete
2. CNR-DT 204 (2006) Guidelines for design, construction and production control of fiber reinforced concrete structures. National Research Council of Italy, Rome
3. IS 1201 to 1220 (1978) Methods for testing tar and bituminous materials
4. IS 1514 (1990) Methods of sampling and test for quick lime and hydrated lime
5. IS 3629 (1988) Structural timber in building—specification
6. IS 9142 (1979) Specification for artificial lightweight aggregates for concrete masonry units
7. RILEM TC 162 TDF (2002) Test and design methods for steel fibre reinforced concrete: Bending test. Mater Struct 35:579–582
8. RILEM TC 162-TDF (2002) Test and design methods for steel fibre reinforced concrete: design of steel fibre reinforced concrete using the σ-w method: principles and applications. Mater Struct 35:262–278

Bibliography

9. American Concrete Institute (1993) ACI 544.4R–88: design considerations for steel fiber reinforced concret
10. ASQ/ANSI/ISO 9000:2015: Quality management systems—fundamentals and vocabulary (E-Standard)
11. ASTM C1688: Standard test method for density and void content of freshly mixed concrete. ASTM International, West Conshohocken, PA
12. CEB-FIP. Model code (2013) First complete draft, vol 1. *Fib* bull 2010 55:1–318
13. EN 14889-1 (Part 1) Fibres for concrete—steel fibres: definitions, specifications and conformity
14. FIB (2010) Model code 2010—first complete draft. International Federation for Structural Concrete, Lausanne
15. IS 73 (2006) Paving bitumen
16. IS 10262 (2009) Guidelines for concrete mix design proportioning
17. IS 1285 (2002) Wrought aluminum and aluminum alloys—extruded round tube and hollow sections for general engineering purposes
18. IS 15388 (2003) Specification for silica fume
19. IS 1599 (1985) Method for bend test (second revision)
20. IS 1608 (2005) Metallic materials—tensile testing at ambient temperature (third revision)
21. IS 1730 (1989) Dimensions for steel plate sheets, strips and flats for general engineering purposes (second revision)
22. IS 1732 (1989) Dimensions for round and square steel bars for structural and general engineering purposes (second revision)
23. IS 1757 (1988) Method for Charpy impact test (V-notch) for metallic material (second revision)
24. IS 1786 (2008) High strength deformed steel bars and wires for concrete reinforcement
25. IS 2062 (2011) Hot rolled medium and high tensile structural steel
26. IS 2386 (1963) Methods of test for aggregates for concrete, Parts 1 to 5
27. IS 2394 (1984) Code of practice for application of lime plaster finish
28. IS 2541 (1991) Code of practice for preparation and use of lime concrete

29. IS 2645 (1975) Specification for integral cement waterproofing compounds
30. IS 3036 (1992) Code of practice for laying lime concrete for a waterproofed roof finish
31. IS 3548 (1988) Code of practice for glazing in buildings
32. IS 3812-1 (2003) Specification for pulverized fuel ash, Part 1: for use as Pozzolana in cement
33. IS 3812-2 (2003) Specification for pulverized fuel ash, Part 2: for use as admixture in cement mortar and concrete
34. IS 383 (1970) Specification for coarse and fine aggregates from natural sources for concrete
35. IS 432-1 (1982) Mild steel and medium tensile steel bars and hard-drawn steel wire for concrete reinforcement, Part 1: mild steel and medium tensile steel bars
36. IS 455 (1989) Portland slag cement—specification
37. IS 456 (2000) Plain and reinforced concrete—code of practice
38. IS 516 (1959) Method of tests for strength of concrete
39. IS 6932-7 (1973) Method of test for building limes, Part VII: determination of compressive and transverse strength
40. IS 712 (1984) Specification for building limes
41. IS 738 (1994) Wrought aluminum alloys, drawn tube for general engineering purposes
42. IS 9103 (1999) Specification for concrete admixtures
43. ISO 13270 (2013) Steel fibres for concrete—definitions and specifications
44. RILEM (2001) RILEM TC 162-TDF—tests and design methods for steel fibre reinforced concrete: uniaxial tension test for steel fibre reinforced concrete. Mater Struct 34:3–6
45. IS 269:1989—Specification for ordinary Portland cement, 33 grade
46. IS 383:1970—Specification for coarse and fine aggregates from natural sources for concrete
47. IS 455:1989 Specification for Portland slag cement
48. IS 456:2000 Code of practice for plain and reinforced concrete
49. IS 457:1957 Code of practice for general construction of plain and reinforced concrete for dams and other massive structures
50. IS 516:1959 Method of test for strength of concrete
51. IS 650:1991 Specification for standard sand for testing of cement
52. IS 1199:1959 Methods of sampling and analysis of concrete
53. IS 1343:1980 Code of practice for prestressed concrete
54. IS 1344:1981 Specification for calcined clay pozzolana
55. IS 1489:1991 Specification for Portland pozzolana cement: Part 1 flyash based
56. IS 1489:1991 Specification for Portland-pozzolana cement: Part 2 (calcined clay based)
57. IS 1727:1967 Methods of test for pozzolanic materials
58. IS 2386:1963 Methods of test for aggregates for concrete: Part 1 (particle size and shape)
59. IS 2386:1963 Methods of test for aggreegates for concrete: Part 2 (estimation of deleterious materials and organic impurities)
60. IS 2386:1963 Methods of test for aggregates for concrete: Part 3 (specific gravity, density, voids, absorption and bulking)
61. IS 2386:1963 Methods of test for aggregates for concrete: Part 4 (mechanical properties)
62. IS 2386:1963 Methods of test for aggregates for concrete: Part 5 (soundness)
63. IS 2386:1963 Methods of test for aggregates for concrete: Part 6 (measuring mortar making properties of fine aggregates)
64. IS 2386:1963 Methods of test for aggregates for concrete: Part 7 (alkali aggregate reactivity)
65. IS 2430:1986 Methods for sampling of aggregates for concrete
66. IS 2502:1963 Code of practice for bending and fixing of bars for concrete reinforcement
67. IS 2645:2003 Integral waterproofing compounds for cement mortar and concrete—specification
68. IS 2770:1967 Methods of testing bond in reinforced concrete: Part 1 pull-out test
69. IS 3085:1965 Method of test for permeability of cement mortar and concrete
70. IS 3370:2009 Code of practice for concrete structures for storage of liquids: Part 1 (general requirements)
71. IS 3370:2009 Code of practice for concrete structures for storage of liquids: Part 2 (reinforced concrete structures)

72. IS 3370:1967 Code of practice for concrete structures for the storage of liquids: Part 3 (prestressed concrete)
73. IS 3370:1967 Code of practice for concrete structures for the storage of liquids: Part 4 (design tables)
74. IS 3466:1988 Specification for masonry cement
75. IS 3535:1986 Methods of sampling hydraulic cement
76. IS 3558:1983 Code of practice for use of immersion vibrators
77. IS 3812:2003 Specification for pulverized fuel ash: Part 1 (for use as pozzolana in cement, cement mortar and concrete)
78. IS 3812:2003 Specification for pulverized fuel ash: Part 2 (for use as admixture in cement mortar and concrete)
79. IS 4031:1996 Methods of physical tests for hydraulic cement: Part 1 (determination of fineness by dry sieving)
80. IS 4031:1999 Methods of physical tests for hydraulic cement: Part 2 (determination of fineness by specific surface by Blaine air permeability method)
81. IS 4031:1988 Methods of physical tests for hydraulic cement: Part 3 (determination of soundness)
82. IS 4031:1988 Methods of physical tests for hydraulic cement: Part 4 (determination of consistency of standard cement paste)
83. IS 4031:1988 Methods of physical tests for hydraulic cement: Part 5 (determination of initial and final setting times)
84. IS 4031:1988 Methods of physical tests for hydraulic cement: Part 6 (determination of compressive strength of hydraulic cement—other than masonry cement)
85. IS 4031:1988 Methods of physical tests for hydraulic cement: Part 7 (determination of compressive strength of masonry cement)
86. IS 4031:1988 Methods of physical tests for hydraulic cement: Part 8 (determination of transverse and compressive strength of plastic mortar using prism)
87. IS 4031:1988 Methods of physical tests for hydraulic cement: Part 9 (determination of heat of hydration)
88. IS 4031:1988 Methods of physical tests for hydraulic cement: Part 10 (determination of drying shrinkage)
89. IS 4031:1988 Methods of physical tests for hydraulic cement: Part 11 (determination of density)
90. IS 4031:1988 Methods of physical tests for hydraulic cement: Part 12 (determination of air content of hydraulic cement mortar)
91. IS 4031:1988 Methods of physical tests for hydraulic cement: Part 13 (measurement of water retentivity of masonry cement)
92. IS 4031:1989 Methods of physical tests for hydraulic cement: Part 14 (determination of false set)
93. IS 4031:1991 Methods of physical test for hydraulic cement: Part 15 (determination of fineness by wet sieving)
94. IS 4032:1985 Method of chemical analysis of hydraulic cement
95. IS 4305:1967 Glossary of terms relating to pozzolana
96. IS 4634:1991 Methods for testing performance of batch-type concrete mixers
97. IS 4845:1968 Definitions and terminology relating to hydraulic
98. IS 4926:2003 Ready mixed concrete—code of practice
99. IS 5512:1983 Specification for flow table for use in tests of hydraulic cements and pozzolanic materials
100. IS 5513:1996 Specification for Vicat apparatus
101. IS 5514:1996 Specification for apparatus used in Le-Chatelier test
102. IS 5515:1983 Specification for compaction factor apparatus
103. IS 5516:1996 Specification for variable flow type air-permeability apparatus (Blaine type)
104. IS 5525:1969 Recommendations for detailing of reinforcement in reinforced concrete works

105. IS 5536:1969 Specification for constant flow type air- permeability apparatus (Lea and Nurse type)
106. IS 5816:1999 Method of test for splitting tensile strength of concrete
107. IS 6452:1989 Specification for high alumina cement for structural use
108. IS 6491:1972 Method of sampling fly ash
109. IS 6909:1990 Specification for supersulfated cement
110. IS 6925:1973 Methods of test for determination of water soluble chlorides in concrete admixtures
111. IS 7246:1974 Recommendations for use of table vibrators for consolidating concrete
112. IS 7320:1974 Specification for concrete slump test apparatus
113. IS 7325:1974 Specification for apparatus for determining constituents of fresh concrete
114. IS 7861:1975 Code of practice for extreme weather concreting: Part 1 (recommended practice for hot weather)
115. IS 7861:1981 Code of practice for extreme weather concreting: Part 2 (recommended practice for cold weather concreting)
116. IS 8041:1990 Specification for rapid hardening Portland cement
117. IS 8042:1989 Specification for white Portland cement
118. IS 8043:1991 Specification for hydrophobic Portland cement
119. IS 8112:1989 Specification for 43 grade ordinary Portland
120. IS 8125:1976 Dimensions and materials of cement rotary kilns, components and auxiliaries (dry process with suspension preheater)
121. IS 8142:1976 Method of test for determining setting time of concrete by penetration resistance
122. IS 8425:1977 Code of practice for determination of specific surface area of powders by air permeability
123. IS 9012:1978 Recommended practices for shotcreting
124. IS 9013:1978 Method of making, curing and determining compressive strength of accelerated cured concrete test specimens
125. IS 9103:1999 Specification for admixtures for concrete
126. IS 9142:1979 Specification for artificial light weight aggregates for concrete masonry units
127. IS 9284:1979 Method of test for abrasion resistance of concrete
128. IS 9376:1979 Specification for apparatus for measuring aggregate crushing value and ten percent fines
129. IS 9377:1979 Specification for apparatus for aggregate impact
130. IS 9399:1979 Specification for apparatus for flexural testing of concrete
131. IS 9459:1980 Specification for apparatus for use in measurement of length change of hardened cement paste, mortar and concrete
132. IS 9799:1981 Specification for pressure meter for determination of air content of freshly mixed concrete
133. IS 10070:1982 Specification for machine for abrasion testing of coarse aggregates
134. IS 10078:1982 Specification for jolting apparatus for testing
135. IS 10079:1982 Specification for cylindrical metal measures for use in tests of aggregates and concrete
136. IS 10080:1982 Specification for vibration machine for casting of standard cement mortar cubes
137. IS 10086:1982 Specification for moulds for use in tests of cement and concrete
138. IS 10262:2009 Guidelines for concrete mix proportioning
139. IS 10510:1983 Specification for Vee-Bee consistometer
140. IS 10850:1984 Specification for apparatus for measurement of water retentivity of masonry cement
141. IS 10890:1984 Specification for planetary mixer used in tests of cement and pozzolana
142. IS 11262:1985 Specification for calorimeter for determination of heat of hydration of hydraulic cement

143. IS 1985:2002 Specification for cylinder measures for determination of air content of hydraulic cement
144. IS 11578:1986 Method for determination of specific surface area of powder and porous particle using low temperature gas absorption techniques
145. IS 11993:1987 Code of practice for use of screed board concrete vibrators
146. IS 12089:1987 Specification for granulated slag for manufacture of Portland slag cement
147. IS 12119:1987 General requirements for pan mixers for concrete
148. IS 12269:1987 Specification for 53 grade ordinary Portland
149. IS 12330:1988 Specification for sulfate resisting Portland
150. IS 12600:1989 Specification for low heat Portland cement
151. IS 12803:1989 Methods of analysis of hydraulic cement by X-ray fluorescence spectrometer
152. IS 12813:1989 Method of analysis of hydraulic cement by atomic absorption spectrophotometer
153. IS 12870:1989 Methods of sampling calcined clay pozzolana
154. IS 13311:1992 Methods of non-destructive testing of concrete: Part 1 (ultrasonic pulse velocity)
155. IS 13311:1992 Methods of non-destructive testing of concrete: Part 2 (rebound hammer)
156. IS 14345:1996 Specification for autoclave apparatus
157. IS 14687:1999 Guidelines for falsework for concrete structures
158. IS 14858: 2000 Requirements for compression testing machine used for testing of concrete and mortar
159. IS 14959: 2001 Method of test for determination of water soluble and acid soluble chlorides in mortar and concrete: Part 1 (fresh mortar and concrete)
160. IS 14959: 2001 Method of test for determination of water soluble and acid soluble chlorides in mortar and concrete: Part 2 (hardened mortar and concrete)
161. IS 15388:2003 Silica fume—specification

Chapter 7
Design of Materials

7.1 Introduction

Design is a plan or set of specifications that any material or the objects made from any such materials must meet to perform satisfactorily within the given set of constraints; usually defined by the requirements of aesthetic, functional use, economic and/or environmental considerations. It is an integral part of any construction project wherein the focus is to get the best out of the available resources at the reasonable costs. Once the architectural/conceptual drawings of any project are finalized, the very next step is to decide the type of materials that will be used therein to construct the facilities. These two steps usually go hand-in-hand, and the architects have to keep in mind the priorities while freezing the things; Sometimes, the type of material dictates the planning and facade of the buildings, and while in some other cases, the planning, etc., become more important and are frozen first and any material that meets the identified objectives can be selected to perform the intended function. Either way, architects and/or engineers have to decide and frame the material and other specifications based upon the project constraints, and the construction process is scheduled accordingly.

In engineering context, the material design usually involves deciding various strengths and durability-related parameters that enable the selected material in any component of structural systems to allow a safe transfer of the forces along a pre-defined load-path. As such, the job of engineers is to decide the comfort zone of materials in which they can be used/ stressed to perform their assigned role satisfactorily. Usually, a set of extensive testing and evaluation exercise is undertaken to fix this zone, especially in case of the new materials or where the available materials are to be used in some new environmental conditions. In all other circumstances, the material parameters are generally prescribed by most of the codes and other similar standard specifications; these are used to decide the safe load carrying capacity of materials and their other design andconstruction related specifications. This chapter attempts to describe the design philosophies and methods that can be

H. Singh, *Structural Materials*, Materials Horizons: From Nature to Nanomaterials,
https://doi.org/10.1007/978-981-16-3211-2_7

used to fix the comfort zone of materials that defines the safe operational limits within which they can be stressed during their use in different members of structural system.

7.2 Design Philosophy

The very first objective of designing materials is that these should not fail or causes any break in the pre-defined load-path in a structural system during their intended service life; the others can be their durability, cost competitiveness and sustainability issues. The only way to ensure compliance of these set objectives is to test the materials and evaluate their response in light of the adopted accepted criterion. The best one will be the case which gives the highest possible value of the strength-to-weight ratio in any given set of the constraints, viz-a-viz: availability, constructability, compatibility with the other materials being used in the project, and design life, etc.

Two design philosophies are commonly used to derive the safe stress values for different materials based upon the test results, namely the infinite-life design and the damage-tolerant design approach. Among these, the infinite-life design is the oldest of the design philosophies and is based on maintaining the stresses in the materials at some fraction below their evaluated ultimate strength. Usually, a suitable factor of safety is used to derive a design value of the stress for any material from its peak load carrying capacity defined in the form of ultimate strength. This type of approach is most applicable when the materials are designed to be stressed in their elastic range only and they display a distinct endurance limit. The material carries the load without developing any form of cracking. It is the most widely practiced approach in the structural designing due to its simplicity and almost zero need of any periodic inspections. The only drawback with this approach is that it produces heavy member sections due to lower stress levels adopted in the design process.

In the second approach, the material is allowed to crack under the prevailing stress conditions, but its maximum width is limited to some pre-set design levels that do not impair the desired functionality of the structural member during its service life. It is a modified form of the previous approach where the design stress level is decided based upon the certain pre-set performance levels instead of some arbitrarily selected value of the factor of safety. By doing so, engineers are able to use a large reserve of the inherent material strength in the design that otherwise remains unused. The material is made to exhibit a higher strength and/or toughness by activating the inherent multiple internal load-paths; the internal switching of various stress transfer mechanisms in some materials (e.g., Composites) makes it possible. The strength and other material parameters are determined from the stress–strain response test data obtained from the laboratory testing of the material specimens. Usually, the probability-based analysis techniques are used to decide the characteristic value of various strength parameters required in structural designing of different members. Analytically, the principles of fracture mechanics can be used to compute the possible position and width of potential cracks at the design strain levels, and accordingly, the stress value can be determined for any subsequent use. The following sections describe and outline the design aspects of different materials.

7.3 Natural Materials

Materials behave differently under the imposed loading conditions depending upon their composition and the microstructure. Natural materials, such as stones, mostly exhibit brittle behavior and suddenly fail to carry loads beyond certain limits depending upon their limiting strain values while the metals and alloys undergo a ductile response before their failure. The composites on the other hand can respond either way; they being designers material can be designed to fail at some pre-defined load levels or to possess a significant strength as well as toughness. For instance, the plain concrete like the stones is a brittle material and if it is reinforced suitably, transforms to the ductile material. While designing materials, the infinite-life approach is generally used to estimate the design stress levels for the natural materials as they have only internal bonding as a sole mode of the stress transfer across them and the chances of internal switching to the other mechanisms are almost non-existent in them. In case of the composites and other materials possessing a significant ductile behavior, the damage-tolerant design approach is better equipped to assess their post-peak strength parameters. Nevertheless, an effective testing and evaluation strategy are a prerequisite to get the desired results. The principles of fracture mechanics and other analytical methods are also helpful to predict the safe stress values of composite materials. Numerous formulations are available to find the safe load carrying capacity of composites like concrete and some other reinforced materials.

7.3.1 Stones

These are the natural materials and are available as stratified or unstratified blocks mined from the stone quarries for various applications. The stratified stones are easy to dress while the unstratified ones are hard and difficult to dress. As such, the design process in this case consists of only selecting suitable form and type of the stone that meets the requirements of a project and subsequently, evaluating their load carrying capacity through laboratory investigations. The stones possessing a minimum compressive strength of 3.5 MPa and having a specific gravity of 2.4 and water absorption of less than 5% are recommended for use as a structural material. Normally, stones to be used as structural material should be dressed in ways that enable their easy lifting and placement by hand during the construction. The length of the stone should not be kept more than three times their height and the breadth on the base shall not be greater than three-fourth of the thickness of the wall nor less than 150 mm. The height of the stone may be kept up to 300 mm. These restrictions are mandated to avoid their sudden failure during the lifting and placement operations and other construction- related activities; they being the brittle material possesses low flexural strengths and suddenly fail when the stress caused by external loading or even their own self-weight exceeds the safe carrying capacity. As a ready reference, a typical set of different strength properties of stones derived from various rocks are given in Table 7.1.

In the absence of test data, the values given in Table 7.1 can be used in the preliminary stage of any design and/or material selection process. However, the actual stone strengths must be determined from the prescribed laboratory investigations (IS 1121). A high value of the factor of safety (usually of 10) is recommended to convert the stone's strength to its safe values. A large variation in the stone composition and the presence of other defects in the rocks necessitate adoption of a high factor of safety in comparison with other materials. Knowing the comfort zone of stones, usually defined by the safe stresses, they can be used to proportion different structural members of buildings.

It is important to note that the design strength of stone masonry will be something different; it is not the same as given in Table 7.1 or derived in the form of safe stress values [7]. A number of factors play their role to give application specific design strength values. The majority among these factors are the mortar composition and the slenderness ratio of a structural element. The detailed design procedure is prescribed in IS 1905 in this regards.

7.3.2 Timber

Like the stones, the timber is also a natural material, but it exhibits some ductility if tested under some bending stress conditions. The presence of organic polymeric material in the form of fibrous structural tissues in timber makes them behave differently in comparison with the stones. As such, the design process involves only selection of an appropriate grade and the class of a timber that fulfills the requirements of any project. The cutting and placement operations control the strength properties. An attempt should be made to cut the wooden logs along the grains/fiber alignments in order to get the maximum possible flexural strength. The defects should be kept to a bare minimum and that too in the permitted zones of the structural members. The design strength values prescribed in IS 3629 should be used to proportion different structural components.

Table 7.1 A set of typical strength properties of different stones

Stone	Compressive strength (MPa)	Flexural strength (MPa)	Modulus of elasticity (GN/m^2)	Tensile strength (MPa)	Specific gravity	Porosity (%)	Coefficient of thermal expansion (10^{-6}/°C)
Igneous rocks							
Basalts	110–335	14–55	60–100	10–30	2.90–3.05	0.3–2.0	2.2–3.4
Dolerites	150–185	8–15	50–75	15–35	2.85–3.05	1.0–2.0	2.5–3.5
Gabbros	120–300	10–55	70–110	15–30	2.80–3.05	0.3–2.7	2.0–3.0
Granites	95–310	9–35	20–60	7–25	2.55–2.65	0.4–2.4	3.7–6.0
Sedimentary rocks							
Limestone	15–255	3–35	10–80	5–25	1.75–2.90	0.3–3.6	1.7–6.8
Sandstones	35–250	5–15	3–80	4–25	1.85–2.75		3.7–6.3
Metamorphic rocks							
Gneisses	50–250	8–20		5–20	2.64–2.35	0.5–0.8	1.3–4.4
Marbles	70–240	4–25		7–20	2.35–3.20	0.6–2.3	2.7–5.1
Quartzite	50–325	8–30		10–30	2.75–3.15	0.3–1.5	2.9–6.0
Slates	70–205	5–20		7–20	2.71–2.90	0.1–4.3	4.5–4.9

Source Author

7.4 Metals and Alloys

Metals are the natural materials, whereas alloys are the man-made produced by mixing two or more metals in their molten state. Both of them find enormous applications in modern day-to-day life due to their very unique characteristics in comparison with many other available materials. As described in the first chapter, the metals are crystalline materials composed of a large number of crystals. The crystals in them may be all alike, in which case the metal is known as single phase, or may be of two different kinds giving a two-phase material.

Atoms in the metals pack themselves in a regular geometric array. The density of packing in any given space is generally limited by the repulsive forces between the adjacent atoms, thereby giving a specific structural layout to the crystal lattice formed in the metals. Actually, this layout is a balance between the cohesive forces of bonding and these repulsive forces; it leads to the formation of three fundamental structural layouts in the metals. This arrangement of atoms in any crystalline lattice structure has a significant effect on the mechanical properties of the metals. For instance, the metals having a body-centered cubic crystalline lattice (e.g., steel) generally exhibit high strength and limited ductility; whereas those with face-centered cubic structure (e.g., copper) possesses lower strength and higher ductility. On the other hand, the metals with the third possibility, namely the hexagonal close-packed, mostly produce metals having brittle characteristics. An important example of the metals having a hexagonal crystalline lattice is zirconium. So, any action that affects the formation of the crystalline lattice in the metals will control their mechanical properties.

Defects are one of the factors that disrupt the regularity of the crystal structure, and as such, these exert significant influence the mechanical properties of metals. These can develop during the stage when the crystals are forming in the metals, or may be created artificially by stressing them after the casting process is over. For instance, the yield strength of many materials can be raised by deforming them beyond their yield point in a certain pattern and operating conditions. The resultant permanent deformation in them destroys the existing regularity of the crystal structure and introduces large numbers of imperfections. These imperfections in turn induce internal stress in the crystalline structure because the atoms have to dislocate themselves from their equilibrium, or more stable positions. The internal stress so induced then opposes any external stress field caused by the applied loads, and therefore, the level of external stress (or load) will have to increase to cause further deformation. This internal process results in the enhancement of the metal yield strength. This technique is in use since long to produce steel bars of higher strengths. Fe415 reinforcing steel bars are one of the examples wherein cold deforming was used to raise the yield strength from 250 to 415 MPa. But due to the changes in the crystalline structure, this process results in reduced ductility and toughness. Another similar process that brings in significant alterations in the lattice structure is thermo-mechanical treatment wherein the steel heated to the specific temperature is cooled in a controlled manner to produce a hard outer layer with a softer core in the bar. In this process, it is possible to increase the yield strength significantly as well as ductility to some extent.

Alloying is another way to improve the material characteristics. It uses the property of metals that enable them to mix together in their liquid state, and thus, alloying generally involves melting of one or more metals before mixing. On cooling, new material forms with altogether new set of properties. The dissolved metal atoms may substitute for the regular metal atoms in the parent crystal lattice, or they may fit into free spaces between the regular metal atoms, thereby leading to modification of the lattice structure. Another big difference it will cause to the lattice structure is the introduction of irregularities because the dissolved alloy atoms are generally of different size than the regular atoms present in the parent metal. Internal stresses are thus set up and, as in the case of plastic deformation; they oppose applied stress and effectively raise the strength of the alloy metal. This process is called as pinning; the larger size of the introduced atoms pin the movement of dislocations in the parent metal grains and leads to improvement in the mechanical properties. It also affects the weldability and machining of the end product. Generally, two phase alloys possess better strength and ductility than the single phase alloys; the different crystal sizes in the two phases therein restrict more effectively the movement of dislocations in the granular structure of the resultant material. For instance, the variations in the carbon content produce a variety of material properties in the case of the plain carbon steel. The iron having a carbon content of less than 0.30% is very ductile in nature, and it is called as low carbon steel. It offers a large range of tensile properties via heat treatment and can be hardened to give high surface hardness with abrasion resistance and at the same time good core ductility. It also possesses good weldability and machining characteristics in comparison with the case when the carbon content in the metal is kept more than 0.3%. If the carbon content is increased in the range 0.30–0.45%, the steel exhibits increased hardness and tensile strength, but it comes with a reduction in the ductility and machining capability. The carbon content of more than 0.45% leads to difficulties at the time of welding process, but the steel in this case possesses very high strength usually around 1000 MPa; albeit with a low elongation of 5–8% only, in comparison with the value of more than 15% possessed by low carbon steel. Carbon should therefore be kept as low as possible to maximize the weldability of the end product and also to get the maximum possible strength and ductility. Minimizing alloying elements to the bare minimum needed to meet the performance requirements of the end product also helps in cost cutting.

It is worth to note that heat treatment is a key step in the whole process to get the desired properties from both the metals and the alloys. It is a specialized process and needs to be completed under the pre-identified controlled production conditions to get a desired rearrangement in the crystal lattice and the distribution of alloying elements (e.g., carbon in the case of the steel) therein to get a better pinning action. Slow cooling rates lead to the coarser microstructure, which reduces the strength; whereas rapidly cooling in water or oil, known as quenching, gives the highest possible strength available from any grade due to small crystal and grain size formed in the process. Cooling in still air called as normalizing provides good strength as well as the ductility. For instance, thermo-mechanical treatment (TMT) process involves an 'in-line' heat treatment process in which hot steel rebars coming out of the last rolling mill stand are rapidly quenched through a series of water jets. Rapid quenching provides intensive cooling of the surface resulting in the rebars having hardened surface with hot core. The rebars are then allowed to cool in ambient conditions. During such slow cooling, the heat released from core tempers the hardened surface while the austenitic core is turned into a ferrite–pearlite aggregate composition. TMT process thus changes the structure of the material to a composite structure of ductile ferrite–pearlite composition with a tough surface rim of tempered martensite providing an optimum combination of the high strength, ductility, bendability and other desirable properties. Yield strength is achieved by tempered martensite rim and the ferrite–pearlite core in the rebars ensures ductility. The variation in mechanical properties is high as compared to air-cooled route due to composite macrostructure/microstructure arising from variability in rolling parameters, such as billet temperature variation, water box parameters, conditions of quenching nozzles and tempering temperature. Achieving annular rim is critical for achieving desired properties. If the quenching is non-uniform, then there is a chance of horse-shoe formation which results in huge variation in properties; it must be avoided by adopting suitable measures.

A number of other elements are generally blended with the hot iron to modify the material properties. Some of these are listed below:

Carbon: It is one of the most effective and economical strengthening elements. Normally, its maximum dosage is limited to 0.25% for getting the ductility response and 0.3% for the non-ductile variants. For the higher strength rebars, such as 500S/Fe650/Fe700, it is limited to 0.32%. Carbon is generally found in two forms, either it is in solid solution in the austenite/ferrite phase or it is found as carbide. Carbon increases the chance of formation of the second phase. However, ductility (elongation), toughness and weldability significantly decrease with an increase in carbon beyond certain values.

Manganese: This element is restricted to 0.30–1.65%. It is useful as a deoxidizer and fixes sulfur in the steel as manganese sulfide to prevent hot embrittlement. The presence of the manganese not only imparts the solid solution strengthening to the ferrite, but also lowers the austenite to ferrite, thereby helps to refine the ferrite grain size.

Silicon: This element like the manganese (Mn) is a very efficient solid solution strengthening element. It is necessary for deoxidation of the steel and as an effective ferrite solid-solution hardening element. For proper deoxidation, the silicon (Si) content may be taken in the range 0.1–0.2% with the restriction that the ratio of Mn/Si should be kept more than 3 for a better castability of the steel in continuous billet caster. If its content becomes too excessive, it results in decarburization and causes extra effort to remove the scales. To avoid these types of problems, the Si content may be kept up to 0.8% and is preferably be 0.6% or less.

Sulfur is inevitable impurity, and its content must be limited to 0.04%; otherwise, it results in a very high inclusion level that deteriorates formability. Phosphorus, like the sulfur, be restricted to 0.04% as the high phosphorus levels lead to reduction in toughness and weldability due to segregation of phosphorus into the grain boundaries.

Aluminum (0.05%, maximum) is effective as a deoxidizer and forms aluminum nitride to prevent austenite from having a larger grain size. In the case of continuous casting without submerged entry nozzle, its content should be restricted to 0.01% only.

Chromium (0.8%, maximum) is very cost-effective alloying element for increasing the hardenability. It has a strong tendency to form hard and stable carbides. This property gives chromium steels the ability to resist softening during tempering and makes them hard and wear-resistant.

Copper (0.5%, maximum) is beneficial to atmospheric corrosion resistance if dosages in amounts exceeding 0.2%. Copper is detrimental to surface quality if present in an amount exceeding 0.5%, but the nano-precipitation at this dosage improves the elevated temperature strength.

Molybdenum (0.45%, maximum) is a potent hardenability agent and retards softening at elevated temperatures and improves the creep resistance. Alloying molybdenum in steel reduces the segregation of phosphorous to grain boundaries, thereby improving resistance to inter-granular fracture. It also improves machinability and resistance to corrosion.

Nickel (0.65%, maximum) alloying increases hardenability, corrosion resistance and the resistance to scaling and toughness especially at low temperatures.

The design of steels and/or alloys thus requires a deep knowledge of various metallurgical aspects of different materials. A number of trials and combinations are needed to arrive at a suitable material composition for obtaining a set of the desirable resultant properties. As a civil engineer, the professionals usually try to use the materials by selecting them from the mechanical properties and other characteristics provided by the producers in the form of the catalogues and other laboratory documents in the form of test reports, performance reports, etc., but by having an insight into the material microstructure and its effect on the response and resultant strength properties, they can better understand the materials and employ them in the construction to get the best possible performance from the built structures. Various BIS codes and other guidelines come handy in this process and an attempt must be made to ensure full compliance of various provisions enshrined in these control documents.

7.5 Polymers

These materials exist naturally as well as can be produced artificially. Natural rubber and jute are typical examples of the natural polymers and find immense applications due to their unique characteristics; whereas PVC is an artificial polymeric material. Unlike the natural polymeric materials, those produced artificially can be tailor designed to possess a variety of properties. These can be of thermoplastic form or can be thermosets. The former type softens upon heating and could be then made to flow when some stress is applied, whereas the latter can be softened only once and would not do so reversibly. At the molecular levels, a polymer consists of long chains of molecules called as '*mers*' joined together by means of a covalent bonding as well as another form of bonds, such as hydrogen bonding, dipole interactions, van der Waals forces and ionic bonds, commonly called as secondary bonding. The major element that leads to the formation of these long chains of molecules is a Carbon atom which due to its tetravalency and catenation action easily formed necessary bonding with other carbon atoms or some other similar elements to get them chemically stabilized.

At the chemistry level, the presence of functional group in the *mers*/participating molecules is a prerequisite for the polymerization to happen. The –OH group on alcohol and –COOH on the acid are the basic functional groups that initiate the reactions. Here, functionality is an important parameter that decides the feasibility of the polymerization and denotes the number of bonds a *mer* can form with other *mers* in a reaction. It is worth to note that reactions between the monofunctional mers cannot lead to polymerization; the presence of difunctional mers, such as diols (HO–R–OH) and diacids (HOOC–R–COOH), is the very first requirement for the polymerization reaction to occur. This type of reaction results in the formation of more difunctional mers; their left end can react with another diacid molecules and its right end with another molecule of participating diol. After each subsequent reaction, the growing molecules still remain difunctional and capable of undergoing further growth. Normally, the presence of all such difunctional groups in the reactant molecules leads to chain polymerization as the new diols or diacids go on joining among each other at the ends of the chains only. The chains so formed are long and flexible in nature; the very first characteristics of polymeric materials.

The molecules with a higher order of functionality (maybe three, four, etc.) leads to other two structural forms of polymeric materials. Branched and cross-linked polymers generally are the result of polymerization caused by the reactants consisting of trifunctional molecules. The change in the polymer morphology causes a tremendous alternation in the material response and strength properties. As the length and frequency of the branches on the polymer chains increase, the probability that the branches so formed to reach from one chain to the other, and finally connecting them together in the three dimensions in the form of cross-linked polymeric chains by means of the primary and/or the secondary bonding also increases. The resultant material is generally very stiff and posses higher strength. It is important to take note that in the linear or branched polymers; only the secondary bonds hold the individual polymer chains together. The consequence is that they easily soften with raising temperature when the energy supplied by the heating exceeds the dissociation energy of the bonding responsible for holding these chains together; the covalent bond has the highest dissociation energy of order of 50–200 kcal/mol; whereas the van der Waals bond is the weakest, having this value of 0.5–2 kcal/mol only. The other bonds that commonly develop among the polymeric chains such as hydrogen bond has this value of 3–7 kcal/mol; dipole interactions (1.5–3 kcal/mol); iconic bond (10–20 kcal/mol). Because of this, the linear and the branched polymers generally result in the formation of thermoplastic whereas the cross-linked polymers leads to thermosets type.

The design of polymeric material therefore involves selection of suitable reactant molecules of adequate functionality that could sustain the polymerization during their synthesis. A thorough knowledge of chemistry is a prerequisite to conceptualize the reactions that lead to the formation of various polymeric materials. The main design objectives of this exercise can be achievement of certain mechanical or optical properties. During the polymerization how different chains organize themselves to control these properties. An ordered, regular chain structure allows the different polymer chains to pack together into an ordered, regular and three-dimensional crystal lattice structures that define the material crystallinity. The materials possessing a higher crystallinity are usually denser and possess more strength in comparison with the amorphous material. The crystalline microstructure if we are able to achieve during the polymerization results in the availability of more number of polymer chains per unit area to bear the applied stress; the long and dense packing also helps to generate cumulatively higher secondary bonding in the material than the case having an amorphous form. For this reason, the polypropylene is fairly hard and rigid plastic than the rubber. Other things being equal, the material designed to have a higher ratio of crystallinity to amorphous phase exhibits more strength, rigidity and stability under high temperatures.

In short, the functionality of reactants used in the production is the only factor that controls all these aspects of polymer behavior. It dictates the overall length of the polymer chains, formation of branching and/or their cross-linking. Generally, the processing parameters and the addition of certain chemicals during or afterward the polymerization control the mechanical, thermal and optical properties of polymers. These are described below.

7.5.1 Changing the Processing Parameter

There are many parameters that can be controlled during the polymerization reaction(s) itself and controlling these results in change in the molecular weight, length and degree of the branching in the polymer chains. The major parameters between them are the types and the concentration of monomers in the reaction; the temperature and the pressure conditions maintained during the synthesis process; the reagent used to initiate the polymerization reaction; the solvent in which the polymer is polymerized and the way they are collected (i.e., are they collected in a manner that promotes linear alignment or not).

As said earlier, any alteration in either polymer lengths or the degree of branching leads to a change in the polymer molecular weight and thus produces a variety of polymers with very distinct mechanical properties, the density and the stretchiness/elasticity characteristics. The polymers with linear chains usually stack together better at the molecular level, which increases the density of the polymer (i.e., more molecules in a given area), as well as their tensile strength (through stronger van der Waals bonding) and exhibits more elasticity; as in all such cases, the polymer molecules can slide over each other easier than otherwise. On the other hand, the polymers with branched chains are less stretchy, less dense and have lower tensile strengths for the opposite reasons.

Changing the length of the polymer is often utilized to change the melting point of the polymer. When the polymer chains are longer, they become more and more tangled and cause them to stick together; it becomes harder to break the bonds between the tangled chains, which increase the material melting point and their density. The reverse scenario is applicable when some lower material melting point is required.

7.5.2 Addition of Other Chemicals

The other ways to change the polymer properties is through cross-linking the polymer chains by means of other chemicals called as plasticizers. The polymers only cross-link when certain cross-linker is added during the polymerization reaction; albeit, under some specific reaction conditions. Cross-linking the polymer chains not only increases the material tensile strength, but it also changes the elastic properties of the polymer. It is important to note that the degree of elasticity of any cross-linked polymer depends only on the polymer type and the cross-linking only assists the molecules to return to their original shape easily than the non-cross-linked polymers. The additives (also referred as plasticiser) change their elastic properties by acting as a molecular lubricant and help the molecules to slide over each other's easily due to reduced friction among different chains. Polymers without plasticizers are much harder and rigid, whereas those which have added plasticizers are more elastic, flexible and softer.

7.6 Composites

Composites are the materials that composed of two or more materials (natural or man-made), blended together with a purpose to improve the material properties. This section describes the design-related issues of some commonly used composites in civil engineering applications.

7.6.1 Mortar

Mortar is one of the oldest building materials in use; the oldest example dates back as far as 10,000 years in Israel. It generally composed of inert material, usually sand, mixed with some binder usually in the presence of water to form a paste that is later used in the masonry construction as a bedding and adhesive to bind and fill the gaps between the bricks or blocks. The mortar is available as a wet ready-to-use type which requires no further water mixing to achieve the required paste consistency and also, as a dry ready-to-use mortar which requires the addition of water before its use. It is categorized based on the type of binder used in its production; different types are listed below.

Cement mortar: When cement is used as binder along with the fine aggregate (such as sand), usually mixed in the ratio of 1:2–1:6; the mortar is called as cement mortar. This provides good strength as well as water resistance.

Aerated cement mortar: In this case, air-entraining chemicals are added to the cement mortar with a purpose to increase its plasticity and workability. The rest of the mix proportions usually remain the same as are used in cement mortar.

Lime mortar: In this case, lime is used as a binder instead of cement. Usually, hydraulic lime is preferred which set relatively fast in comparison with the non-hydraulic lime. The major problem with this type of mortar is its slow setting and low strengths in comparison with the cement- based mortars.

Gauged mortar: It is a mixture of lime, cement and sand taken in certain proportions. It has goodness of both lime- and cement- based mortar; possesses plasticity of lime and the strength characteristics of cement to the mix and also, it results in the cost reduction.

Surkhi mortar: In this type, usually Surkhi (finely powdered burnt clay) is used in place of the sand along with the lime or cement as its binder. It provides more strength than the case when sand is used as filler/inert material.

The material composition of different mortars is given in Table 7.2. In the table, A, B and C denote eminently hydraulic lime, semi-hydraulic lime and fat lime, respectively, as are specified in relevant Indian Standards. The tabulated data can be used to decide the mortar grade needed in the construction work depending upon the strength required to transfer the load through the given type of masonry. Nevertheless, the strength of bricks or blocks used in the construction significantly controls the strength of the composite construction due to their lower strength than that of the mortar. For that purpose, the strength decided from the results of a standard prism test should be used in the design of masonry works.

In addition to the strength, constructability using the mortar is an important aspect that should not be ignored while deciding a suitable type of mix. Generally, the water retention and air content are the vital parameters of any mix that control the strength as well as the ease of handling them during their application. Suitable measures (maybe by means of laboratory studies) must be taken to ensure compliance of the acceptance criterion in this regards; mortar should be able to retain a

Table 7.2 Mortar grades and their compressive strength

Mortar grade	Cement	Lime	Lime-pozzolana mix	Pozzolana	Sand	Compressive strength (MPa)
H1	1	0	0	0	3	10
H2	1	0	0	0	4	7.5
	1	0	0	0	4.5	6
M1	1	0	0	0	5	5
	1	1 (B or C)	0	0	6	3
	0	0	1 (LP-40)	0	1.5	3
M2	1	0	0	0	6	3
	1	2 (B)	0	0	9	2
	0	1 (A)	0	0	2	2
	0	1 (B)	0	1	1	2
	0	1 (B or C)	0	2	0	2
	0	0	1 (LP-40)	0	1.75	2
M3	1	0	0	0	7	1.5
	1	3 (B)	0	0	12	1.5
	0	1 (A)	0	0	3	1.5
	0	1 (B)	0	2	1	1.5
	0	1 (B or C)	0	3	0	1.5
	0	0	1 (LP-40)	0	2	1.5
L1	1	0	0	0	8	0.7
	0	1 (B)	0	1	2	0.7
	0	1 (C or B)	0	2	1	0.7
	0	0	1 (LP-40)	0	2.5	0.7
	0	0	1 (LP-20)	0	1.5	0.7
L2	0	1 (B)	0	0	3	0.5
	0	1 (C or B)	0	1	2	0.5
	0	0	1 (LP-7)	0	1.5	0.5

Source https://archive.org/details/gov.in.is.1905.1987

minimum of 75% of the water to help completion of the hydration reactions, and the maximum air content should be limited to 12% to get an optimum strength as well as mix consistency [10].

7.6.2 Concrete

It is a versatile man-made material that can be designed to exhibit brittle response as well as ductile one. The plain concrete is usually used in situations where the compression stresses dominate and the tensile stresses if any remains well within the safe tensile capacity of the material. Reinforcing the concrete either by means of conventional continuous rebars in the tension zone of members or by mixing short fibers in the concrete at the time of its production improves the brittle characteristics it possesses otherwise. It also enhances the material toughness as well as other mechanical properties.

By controlling the material microstructure and the extent of the reinforcement used in the member, the concrete can be designed to exhibit the desired strength properties. The material composition should be selected in a manner to give the highest possible density within the other field constraints, such as workability, flowability and the segregation resistance. This process needs a number of trials to arrive at the best possible mix proportions that meet all the design objectives simultaneously. IS 10262 prescribes a concrete mix design method along with a few design examples in this regards. It usually starts with the identification of requirements defined by means of characteristic compressive strength and other constraints. From the required quality control level at the construction site, the desired characteristic strength (f_{ck}) is converted to the target concrete strength (f_t) that we aim to achieve under the laboratory conditions; the relation between these two strength parameters is given in Eq. 7.1. This condition will ensure that at least 95% test results from the sample would possess the desired compressive strength.

$$f_t = f_{ck} + 1.64\sigma \tag{7.1}$$

The standard deviation (σ) needed in Eq. 7.1 should be determined from the actual field tests; however, in the absence of such test data, it can be taken as 4 MPa for concrete grades up to M20 and 5 MPa for other the concrete grades up to M55 to start the mix design process, but this assumed value must be compared with the actual value as and when the test data is gathered and the necessary course corrections should be made to adjust the mix proportions. In cases, where a regular checking of the constituent materials and controlled addition of water into the mix not possible; the above-mentioned standard deviation values should be increased by 1 MPa to take into account the higher variations expected in such a scenario.

The cement content and water–cement ratio (including supplementary cementitious materials) that should be ensured in the mix must conform to the prescribed values given in Table 7.3; these are a bare minimum prescription to ensure good durability of the hardened concrete in the respective environmental exposure conditions [5].

For angular coarse aggregate and 25–50 mm slump range, the quantity of water needed per unit volume of concrete may be estimated from Table 7.4. It can be reduced by 10 kg for sub-angular aggregates, 20 kg for gravels with some crushed particles and 25 kg for the rounded gravels to produce the same level of workability; around 3% extra water will be needed for every additional 25 mm slump over and above the prescribed range.

The procedure starts with some assumed value of coarse aggregate size (as per availability in the nearby area or prescribed in the control documents) and adoption of suitable water content from Table 7.4. The water–cement ratio in the trail mix can be adopted from the past experience. These two values give the cement content

Table 7.3 Prescribed values of cement, water–cement ratio for different exposure conditions

Environmental exposure conditions		Plain concrete			Reinforced concrete		
Category	Concrete surface conditions	Minimum cement content (kg/m^3)	Maximum water–cement ratio	Minimum concrete grade	Minimum cement content (kg/m^3)	Maximum water–cement ratio	Minimum concrete grade
Mild	Protected against weather or aggressive conditions	220	0.60	–	300	0.50	M20
Moderate	Sheltered from severe rain or freezing while wet and from saturated salt air in coastal area; exposed to condensation; underwater; Buried under non-aggressive soil/ground water	240	0.60	M15	300	0.50	M25
Severe	Exposed to severe rain, alternate wetting and drying; occasional freezing while wet or severe condensation; immersed in seawater	250	0.50	M20	320	0.40	M30
Very severe	Exposed to sea water; corrosive fumes or severe freezing conditions while wet; in contact with or buried under aggressive sub-soil/ground water	260	0.45	M20	340	0.45	M35
Extreme	In tidal zone; in direct contact with liquid or solid aggressive chemicals	280	0.40	M25	360	0.40	M40

Source https://archive.org/details/gov.in.is.456.2000

Table 7.4 Prescribed values of cement, water–cement ratio for different exposure conditions

Nominal maximum aggregate size (mm)	Maximum water content (kg/m^3)	Aggregate ratio defining the volume of coarse aggregate per unit volume of total aggregates in the mix (for different zones of fine aggregates and the water–cement ratio of 0.5)			
		Zone-IV	Zone-III	Zone-II	Zone-I
10	208	0.50	0.48	0.46	0.44
20	186	0.66	0.64	0.62	0.60
40	165	0.75	0.73	0.71	0.69

Source https://archive.org/details/gov.in.is.383.2016

required to achieve the desired water–cement ratio for a given coarse aggregate size and the concrete workability. Nevertheless, the values so obtained must conform to the prescribed values given in Table 7.3. Depending upon the gradation zone of available fine aggregates, the quantity of coarse aggregates required to attain the maximum possible density of the concrete mix can be estimated from the aggregate-ratio given in Table 7.4 for the water–cement ratio of 0.5. For any other value of water–cement ratio, the aggregate ratio may be adjusted at a rate of ± 0.01 for every ∓ 0.05 changes in the value of the water–cement ratio adopted in the mix design. Similarly, the water content should be adjusted if the adopted water–cement ratio is different from the value used to arrive at the values given in the table; it should be done a rate of $\pm 3\%$ of the adopted value from Table 7.4 for every ∓ 0.05 changes in the value of the water–cement ratio considered in the mix design. Alternatively, it can be taken as +3% of the adopted value (see, Table 7.4) for every 25 mm slump over 50 mm, if the slump is being used as a control parameter. The coarse aggregate content may be reduced up to 10% in the case of self-consolidating concrete and then, the content of the cement and fine aggregates in the mix should be worked out afresh to attain the desired value of the compressive strength and the mix flowability/slump value.

The total quantity of aggregates (**a**) in the mix is estimated from the known absolute volume of cementitious material (**c**), water (**w**) and the chemical admixture (**s**) using Eq. 7.2. The value (**a**) so obtained is divided into coarse and fine aggregate fractions (by volume) by multiplying it with the aggregate ratio already determined from Table 7.2, which if needed can be converted subsequently into mass.

$$a = \left(1 - \frac{c}{1000 s_c} - \frac{w}{1000 s_w} - \frac{s}{1000 s_s}\right) \text{m}^3 \tag{7.2}$$

The parameters s_c, s_w and s_s are the respective values of specific gravities of the cement, the water and the superplasticizers taken in the trial mix. The concrete mix proportions along with the admixture dosage are first decided analytically [as is prescribed above, using Eqs. 7.1 and 7.2] to achieve the target strength and then, a number of trials are conducted to see attainment of the accepted criterion; any trial proportion that meets all the design objectives is taken as a final design mix. This part of the design is a very important component to ensure that the calculated mix proportions meets the acceptance criterion defined with respect to its fresh properties as well as strength properties. The trial mix should carefully be observed for a freedom from segregation and bleeding and its other finishing characteristics. In case of the measured workability coming different from the stipulated value in the mix, the water and/or admixture content may be adjusted suitably and new trail mix should be prepared with the new mix proportions, determined from the new value of the water content keeping the water–cement ratio in the mix the same. Subsequently, the strength properties should be evaluated once the mix meets the requirements in its fresh state.

The standard cubes cast using this mix will exhibit the required characteristic compressive strength (f_{ck}) within the range defined by the acceptance criterion. The tensile strength (in the form of a modulus of rupture) and the elastic modulus associated with this concrete grade can be determined from the expressions: $0.7\sqrt{f_{ck}}$ and $5000\sqrt{f_{ck}}$, respectively (f_{ck} in MPa). Alternatively, the concrete strength/grade needed to achieve the desired tensile strength and/or elastic modulus can be determined from these expressions and then, its mix proportions can be later worked out to determine the quantity of the concrete constituent materials (usually expressed as a volumetric ratio) required to produce the said concrete. All other concrete properties, such as shear strength, shrinkage and creep behavior depends upon the concrete mix proportions finalized for any subsequent use. As such, it is very important to remain careful in choosing the right size and shape of aggregates, cement content and the dosage of admixtures while working out the mix proportions. Like the tensile strength, the toughness of plain concrete is non-significant and it is normally ignored in the structural design calculations. However, concrete does possess some shear strength (though small in magnitude) being contributed by its little tensile strength.

In case of the special (unreinforced) concretes, such as lightweight concrete, permeable concrete or self-consolidating concrete; the procedure prescribed in Chap. 6 should be followed to get these additional characteristics. The mix proportions in all such cases should be determined to meet the specific purpose. The testing program for such type of special concretes needs extensive experimental investigations to ensure that the concrete possesses the desired strength as well as other properties for which these are being produced.

Example: A concrete is required for casting a retaining wall along some drain carrying waste water from industrial units. Design a suitable concrete mix for the following field constraints.

Cement type	OPC 43 grade conforming to IS 8112
Maximum nominal size of aggregate	20 mm
Type of aggregate	Crushed angular aggregate
Workability	100 mm slump

(continued)

(continued)

Cement type	OPC 43 grade conforming to IS 8112
Exposure condition	Severe
Method of concrete placing	Pumping
Degree of supervision	Good
Chemical admixture type	Superplasticizer having specific gravity of 1.15
Maximum cement content	450 kg/m^3

Source Author

The specified exposure condition in this case requires a minimum of M30 concrete grade to ensure an adequate durability level under the prevailing conditions. The experimental testing indicated that the specific gravity of the cement, fine and coarse aggregates is 3.1, 2.74 and 2.74, respectively. The free moisture in the fine and coarse aggregates is found be zero; whereas their water absorption is calculated as 0.5 and 1%, respectively. The sieve analysis data of the fine aggregates revealed that they belong to the zone-I; whereas the test data of available coarse aggregates shows that these are unfit for the use if used as single size aggregates, so coarse aggregates from some other source are procured and is planned to use them by mixing in certain proportions (say, 60% and 40%, respectively) in order to get the desirable gradation. Following table illustrates the mixing process to ensure that coarse aggregates after their mixing meet the gradation criterion of IS 383; it is given in the last column of the table:

IS Sieve sizes, mm	Analysis of coarse aggregate fraction		Percentage of different fractions		
	Source-1	Source-2	60% of source-1	40% of source-2	Combined mix
20	100	100	60	40	100
10	0	71.5	0	28.6	28.6
4.75	0	9.75	0	3.9	3.9
2.36	0	0	0	0	0

Source Author

The required target strength of the concrete mix $f_t = 30 + 1.64 \times 5 = 38.25$ MPa.

Water–cement ratio: Any value can be taken depending upon the past experiences and other success stories, but conforming to the maximum value given in Table 7.3. Let it be 0.35 < 0.4 required for severe environmental conditions.

Water content: The maximum water content can be taken from Table 7.4. For a water–cement ratio of 0.5 and 20 mm aggregate size, this value comes as 186.

This value needs to be modified as the water–cement ratio selected for the trial mix is less than the value specified for the tabulated data (i.e., 0.5).

The modified water content $= 186 + \left\{\frac{(0.5-0.35)}{0.05}\right\} 0.03 \times 186 = 202.74$, but as some superplasticizers is planned to be used during the construction, the water content can further be reduced by 20% in this case. The actual value of the reduction, however, should come from the laboratory investigations; if the study so conducted permits a higher reduction to the desired level of the workability it should be used. Say, this value comes as 25% for an admixture dosage of 2% of the cement mass; the actual water content of 202.74 × (1 − 0.25) = 152 kg/m^3 may therefore be used in the trial mix.

Cement content: Knowing the water–cement ratio (= 0.35) and the water content (= 152 kg/m^3) in the mix, the quantity of the cement required to achieve these values can be determined as 434 kg/m^3 (= 152/0.35). This value must be from the range specified in Table 7.3; it is more than 320 kg/m^3 and less than the imposed maximum limit of 450 kg/m^3; therefore, it is okay to be considered in the subsequent calculations.

Aggregates content: The volume of the coarse aggregate per unit volume of mix can be taken from Table 7.4; this value is obtained as 0.6 corresponding to 20 mm size aggregates and the zone-I of the fine aggregate being used in the mix, for a water–cement ratio of 0.5. In the present case, the water–cement ratio is 0.35. Therefore, the volume of the coarse aggregate is required to be increased depending upon the fine aggregate content. As the water–cement ratio in this case is lower by 0.15 (= 0.5 – 0.35) from the specified value of 0.5, the proportion of the volume of the coarse aggregate should therefore to be increased by 0.03 (= 0.15 × 0.01/0.05). Therefore, the corrected proportion of volume of coarse aggregate for the water–cement ratio of 0.35 is 0.63 (= 0.6 + 0.03).

As the concrete at the site is proposed to be delivered by a pump, the coarse aggregates content in the mix should be reduced by 10% to make flow through the pipeline easy. The volume of the coarse aggregate in the unit volume of the mix, therefore, comes out as 0.567 (= 0.63 × 0.9), and the volume of the fine aggregates required is 0.433 (= 1 – 0.567).

The volume of the aggregates (**a**) in the mix can be determined from Eq. 7.2, as follows:

$$\boldsymbol{a} = \left(1 - \frac{434}{1000 \times 3.1} - \frac{152}{1000 \times 1} - \frac{0.02 \times 434}{1000 \times 1.15}\right) = 0.7\ \text{m}^3$$

Therefore, the mass of the coarse aggregates = **a** × volume of coarse aggregate in the unit volume of the mix × their specific gravity × 1000 = 0.7 × 0.567 × 2.74 × 1000 = 1087.51 kg.

And, the mass of the fine aggregates = **a** × volume of fine aggregate in the unit volume of the mix × their specific gravity × 1000 = 0.7 × 0.433 × 2.74 × 1000 = 830.49 kg.

The first trial can be done with the concrete mix composed of Cement, fine aggregates, coarse aggregates, water and admixture taken in proportions of 434.0: 830.5: 1087.5: 152.0: 8.7 (in kg/m^3), respectively. The concrete sample should be critically evacuated for its desired fresh properties; if found okay, the standard cubes may be cast to check the strength properties and see the compliance of the acceptance criterion, else some changes in the mix proportions should be made, especially in the water content and/or the ratio of fine to coarse aggregates and a new trial mix should accordingly be evaluated in the laboratory again till it meet the design objectives defined in the form of an acceptance criterion. The purpose of mix design has been just to arrive at the trial value based on scientific principles. If some mix proportions are known from the past experiences/earlier success stories, these can be adopted directly as a first trial mix to see the compliance of the acceptance criterion; the necessary modifications can be made in the mix if some non-compliance is noticed.

The concrete reinforced suitably behaves entirely in a different manner. Unlike the plain concrete, it posses adequate bending resistance, shear capacity and toughness to enable a safe transfer of the applied load through the structural members satisfactorily to the supports or some other point of interests in the structural system. The quantity of the reinforcement and its placement in the member needed to accomplish this task is an important part of the design process. For the reinforced rectangular concrete beams (of size width **b** and effective depth **d**) containing conventional longitudinal tension steel ($\mathbf{A_{st}}$), the design shear capacity (factored) can be computed from the tabulated data [5]; see Table 7.5.

Table 7.5 Design shear strength of RC beams having conventional longitudinal tension reinforcement

$\rho_t = \frac{A_{st}}{bd}$	Ultimate concrete shear strength, MPa for different concrete grades					
	M15	M20	M25	M30	M35	$\geq$M40
$\leq$0.15	0.28	0.28	0.29	0.29	0.29	0.30
0.25	0.35	0.36	0.36	0.37	0.37	0.38
0.50	0.46	0.48	0.49	0.50	0.50	0.51
0.75	0.54	0.56	0.$1	0.59	0.59	0.60
1.00	0.60	0.62	0.64	0.66	0.61	0.68
1.25	0.64	0.61	0.70	0.71	0.73	0.74
1.50	0.68	0.72	0.74	0.76	0.78	0.79
1.75	0.71	0.75	0.78	0.80	0.82	0.84
2.00	0.77	0.79	0.82	0.84	0.86	0.88
2.25	0.77	0.81	0.85	0.88	0.90	0.92
2.50	0.77	0.82	0.88	0.91	0.93	0.95
2.75	0.77	0.82	0.90	0.94	0.96	0.98
3.00	0.77	0.82	0.92	0.96	0.99	1.01

Source https://archive.org/details/gov.in.is.456.2000

It is important for the concrete beam to have some minimum amount of tension reinforcement in the section to posses certain minimum shear resistance. This percentage reinforcement in the beam section is specified as $0.85/f_y$ by most of the design codes, corresponding to shear strength of 0.29 MPa. The presence of the tension steel in the beam ensures that the section is in a position to mobilize all the five fundamental stress transfer modes in the concrete as the load approaches the ultimate capacity. Therefore the steel provided in the tension zone of the beam must be adequately anchored along its length, especially the end regions and/or point of the contra-flexure to develop the required tensile force and dowel action. The prevailing stress level in the rebars decides the crack width that would develop in the tension zone of the section which in turn activates the aggregate interlock along the shear plane; the finer the cracks the more effective be the aggregate interlock in the concrete member.

The longitudinal steel in the section also enables the section to possess flexural strength. The rebars in the section develop tension and compression force depending upon their position in the section with respect to the neutral-axis and generate internal couple along with the compressive force being developed in the concrete in the beam compression zone to resist the moment caused by the external loading.

Almost similar things occur in the concrete beams reinforced internally by means of the short, discrete steel fibers (having length $\boldsymbol{l}$ and diameter $\boldsymbol{d}$); these are mixed in the concrete at the time of its production and remain present throughout the concrete mass in the beam in its hardened state. However, these are most effective only in the beam tension zone where they because of the bridging action arrest the formation of the cracks (both flexural and shear) and their subsequent widening as the load approaches its ultimate state. During this process, the fibers can either pull out or fracture under the prevailing tensile stress in the beam section. However, the collapse can be ensured to occur in the more desirable ‘pullout’ failure mode by an appropriate selection of fiber parameters, namely the fiber aspect ratio and its volume fraction. It will occur if the fiber aspect ratio ($\boldsymbol{l/d}$) is kept less than the critical value given in Eq. 7.3 and the fiber dosage ($\boldsymbol{V_f}$) is kept more than the value given in Eq. 7.4, where $\boldsymbol{f_y}$ and $\boldsymbol{f_{ck}}$ are the yield strength of the steel fibers and the concrete compressive strength, respectively [17].

$$\left(\frac{l}{d}\right)_c = \frac{0.58f_y}{f_s\sqrt{f_{ck}}} \tag{7.3}$$

$$V_f \geq 4\frac{\sqrt{f_{ck}}}{f_s f_y} \tag{7.4}$$

In the above expressions, the constant ($\boldsymbol{f_s}$) denotes the effect of the shape of fibers on the behavior of steel fiber- reinforced concrete. The effect is expressed in the form a multiplying constant, known as fiber-shape factor; it can be taken 2, 3 for the hooked-end fibers and the wavy fibers, respectively, used in the concrete production.

The residual tensile strength (**σ_t**) of steel fiber- reinforced concrete (SFRC) in all such cases can be computed analytically using Eq. 7.5. It needs the value of the adopted fiber aspect ratio and its volume fraction along with the concrete compressive strength as input parameters. The strength depicted by Eq. 7.5a is for the case when the specimen fails in the fiber-pullout mode; whereas Eq. 7.5b estimates the material strength when the fibers fracture because of their higher anchorage into the concrete than is required otherwise for the pullout mode.

$$\sigma_t = \begin{cases} 5000\varepsilon\sqrt{f_{ck}} & \varepsilon \leq 0.00014 \\ 0.3f_s\sqrt{f_{ck}}V_f\frac{l}{d} & 0.00014 < \varepsilon \leq \varepsilon_{t(=0.015)} \end{cases} \quad \text{For } \frac{l}{d} < \left(\frac{l}{d}\right)_c \tag{7.5a}$$

$$\sigma_t = \begin{cases} 5000\varepsilon\sqrt{f_{ck}} & \varepsilon \leq 0.00014 \\ 0.87f_yV_f & 0.00014 < \varepsilon \leq \varepsilon_{t(=0.015)} \end{cases} \quad \text{For } \frac{l}{d} < \left(\frac{l}{d}\right)_c \tag{7.5b}$$

The values of the fiber parameters (**V_f** and ***l/d***) that define the transition of the post-cracking behavior of SFRC from a 'strain-softening' to the 'strain-hardening' can be computed from Eq. 7.6.

$$V_f\frac{l}{d} \leq \frac{2.33}{f_s} \tag{7.6}$$

The expressions 7.5 and 7.6 are applicable for the concrete prepared by mixing short and plain fibers at the time of its mixing. If the value of the volume fraction (**V_f**) or the fiber aspect ratio (***l/d***) to be mixed in the concrete mix is kept less than the value computed from Eq. 7.6, the SFRC will exhibit a strain-softening response in its post-cracking range; otherwise, the material would exhibit a strain hardening response.

Example: The M40 concrete mix proportioned in the previous example needs to have a post-cracking tensile strength of 4 MPa and exhibit a strain-hardening response. Design a suitable fiber dosage to have a concrete of these characteristics. Hooked-end fibers are to be used in the concrete production. The yield strength of fibers is 1250 MPa.

The fiber shape factor (***f_s***) for the hooked-end fibers is 2; the fiber dosage needed to get the desired value of material residual tensile strength can be computed from Eq. 7.5a; it is given below:

$$4 = 0.3 \times 2 \times \sqrt{40}V_f\frac{l}{d} \Rightarrow V_f\frac{l}{d} = 1.05$$

The above equation provides a number of options to decide an appropriate combination of the fiber parameters needed to achieve the desired tensile strength; the final value depends upon the achievement of required workability and other characteristics of that mix proportions. Table 7.6 lists all such possible combinations to achieve the desired value of residual tensile strength. Equations 7.3 and 7.4

Table 7.6 Design options to select an appropriate combination of fiber parameters

Aspect ratio, l/*d*	Fiber dosage		Min. volume fraction, V_f (%) needed to ensure strain hardening response (see, Eq. 7.6)
	Volume fraction, V_f (%)	Weight of the steel fibers (kg/m^3)	
40	2.63	206	2.91
50	2.10	165	2.33
55	1.92	150	2.12

Source Author

can be used to compute the critical aspect ratio and the fiber dosage required to activate the fiber pullout failure mode at the ultimate state; these are given below:

$$\left(\frac{l}{d}\right)_c = \frac{0.58 \times 1250}{2\sqrt{40}} = 57.31$$

$$V_f \geq 4\frac{\sqrt{40}}{2 \times 1250} = 1.01\%$$

The fiber aspect ratio should be kept less than its critical value (given above), and any fiber dosage to be used in the mix must be kept more than 1.01%.

The above table indicates that the SFRC specimen will exhibit a strain-softening response at all values of the fiber aspect ratio as the volume fraction required at that level is less than the value needed to prevent the strain softening (compare the second and the last column of the table). As the presence of fibers in the mix tends to reduce the workability, so trials should be conducted to see the attainment of the required slump or flowability and other listed acceptance criterion before finalizing the mix for any subsequent use. Generally, it is difficult to mix fiber if its dosage is kept more than 2–2.5%. In this case, if the fiber aspect ratio is kept more than 55, chances are that the fibers may fracture at the ultimate load. So, if someone is interested in getting the strain hardening response, the fiber dosage has to be kept more than the value indicated in the last column of the table; it can be conveniently done for the aspect ratio of 50 or 55. Nevertheless, a laboratory study is a pre-requisite to get the desired strength and/or response as well as constructability; it should be done for all combinations of the fiber aspect ratio and the fiber dosage selected in Table 7.6.

A number of empirical relations are available to estimate the material strength properties; the Annexure lists various available empirical expressions that can be used to estimate shear strength of the rectangular beams cast using the concrete containing steel fibers. It can also be computed analytically. The fiber index ($\boldsymbol{\beta}$) is a major influential factor among many others that control the shear strength of fiber-reinforced concrete (SFRC) beams; it is defined as a ratio of the residual design tensile strength of the concrete to its 28 day compressive strength ($= \sigma_t/f_{ck}$). The word 'design' in the statement signifies the value of the concrete tensile strength corresponding to any desired crack width considered in the design process; it can be a crack mouth opening displacement (CMOD) of 3.5 mm as required by the RELIM [15], or a CMOD of 1.5 mm as is mandated by the CNR-DT-204 [6] guidelines.

The addition of steel fibers into the concrete mix enhances shear capacity of the beams cast using SFRC. For a concrete beam section having a width (**B**) and an overall depth (**D**) produced by mixing suitable steel fibers, having fiber parameters (***l/d*** and $\boldsymbol{V_f}$), shear strength ($\boldsymbol{\tau_{cf}}$) being contributed by the concrete and the steel fibers in the section can be computed from the following Eq. 7.7 [18].

$$\frac{\tau_{cf}}{f_c} = \frac{k\beta^{0.94}}{3\left(\frac{a}{D}\right)^2} \tag{7.7}$$

Equation 7.7 indicates that the shear strength (τ_{cf}) of SFRC beam section depends only on the residual tensile strength of the concrete, expressed in the form of the fiber index ($\boldsymbol{\beta}$) and the shear span-to-depth ratio (***a/D***). The parameter (k) in the expression can be taken as $1.4/\beta\sqrt{f_c}$. The presence of conventional reinforcing bars in the section activates the dowel action and the section exhibits higher shear strength than the case when it is reinforced using steel fibers alone. The shear strength ($\boldsymbol{\tau_c}$) at the onset of the first shear crack in the section, suitably reinforced using conventional longitudinal tension steel (having a percentage of $\boldsymbol{\rho_t}$), can be computed using Eq. 7.8. The term $0.3\sqrt{\rho_t f_c}$ in the expression depicts the contribution from the dowel action that the conventional reinforcement in the section would provide. It is important to note that the effect of the shear span-to-depth ratio (***a/D***) diminishes fast on the beam shear strength at the higher values (>3.5); however, it is the most influential parameter controlling the section shear strength, especially for the values of ***a/D*** up to 2.5.

$$\tau_c = \frac{kf_c\beta^{0.94}}{3\left(\frac{a}{D}\right)^2} + 0.3\sqrt{\rho_t f_c} \tag{7.8}$$

However, when the beam is further loaded beyond the value that causes initiation of diagonal cracks in the member (assuming the beam to be strong enough to resist the flexural stresses safely), the aggregates and the fibers protruding out from the surface of some shear crack start contributing toward the shear strength in the form of the bridging action of the fibers present therein and the aggregate-interlocking. Equation 7.9 gives the value of shear strength ($\boldsymbol{\tau_{ai}}$) that gets mobilized along the shear crack because of the fiber bridging action.

$$\tau_{ai} = \frac{(k+1)}{5} f_c\left(\beta - 0.1\alpha\beta^{0.3}\right) \tag{7.9}$$

In the above equation, the parameter α is a factor that considers the effect of the longitudinal tension reinforcement (ρ_t) on the crack formation, and it is equal to

Table 7.7 Ultimate shear strength of SFRC slender rectangular beams ($a/D \geq 3.5$)

Fiber index (β)	Ultimate concrete shear strength, MPa for different concrete grades			
	M20	M25	M30	M35
0.05	0.47	0.53	0.59	0.64
0.10	1.02	1.16	1.30	1.43
0.15	1.34	1.55	1.74	1.93
0.20	1.59	1.86	2.12	2.36
0.25	1.84	2.15	2.45	2.75
0.30	2.06	2.43	2.78	3.12

Source Author

0.25 for $\omega_t \geq 0.1$ and is $(1 - 7.5\omega)$ for $\omega_t < 0.1$, where ω_t is the tension reinforcement index of the section $\left(= \rho_t f_y / f_c\right)$. The ultimate shear strength (τ_u) of steel fiber- reinforced concrete beam section can be obtained by adding Eqs. 7.8 and 7.9. Based upon these equations, a typical set of the ultimate concrete beam shear strength as a function of the concrete grade (without any longitudinal tension steel) and the fiber index values is given in Table 7.7. The corresponding design value of the beam shear strength can be determined by using a partial safety factor of 1.5 or by using a factor as mandated by design guidelines.

The value of the fiber index (***β***) needed to provide the desired concrete shear resistance can be determined easily from the tabulated data. It indicates that for a given concrete grade, the shear strength of any SFRC beam can be increased by using the steel fibers having a higher aspect ratio (***l/d***) and/or higher fiber dosage (***V***) in the concrete mix. Their final selection, however, always comes from the workability and placability requirements of the concrete mix at the site.

The bending strength of concrete rectangular beams (see, Fig. 7.1) can be computed from a normalized expression given in Eq. 7.10. The beam section is assumed to be reinforced using the conventional reinforcement on its compression and tension zone, expressed by means of their respective reinforcement indexes (ω_c and ω_t) as well as the steel fibers mixed in the concrete at the time of its production. The value of the factor (ξ) in the expression should be taken as unity for the SFRC;

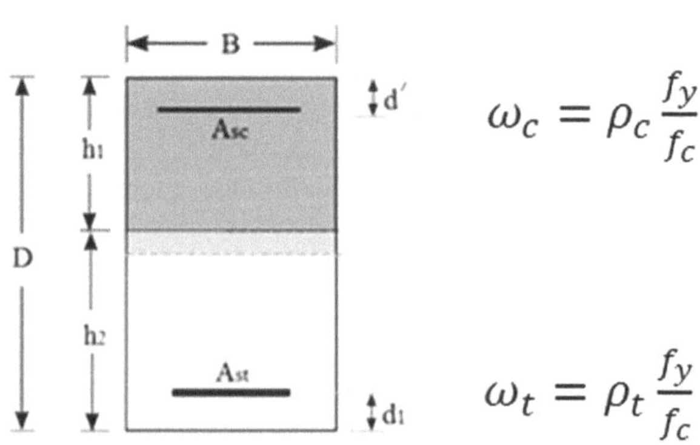

Fig. 7.1 A beam schematic diagram, reinforced using conventional reinforcement as well as steel fibers. *Source* Author

otherwise, it is 0.86 for the beam section being cast in conventional reinforced concrete without steel fibers. Various notations used in Eq. 7.10 are schematically shown in Fig. 7.1.

$$\frac{M_u}{f_{ck}BD^2} = 0.24\xi\left(\frac{h_1}{D}\right)^2 + \left\{0.5\beta + \gamma_t\omega_t\left(\frac{D}{h_2}\right)\left(1 - \frac{d_1}{D}\frac{D}{h_2}\right)\right\}\left(\frac{h_2}{D}\right)^2 + \gamma_c\omega_c\left(1 - \frac{d'}{D}\right) \tag{7.10}$$

In the expression, h_1/D and h_2/D define the position of neutral section in the beam section. Using the following expressions, these two values can be computed from the known value of the reinforcement index (ω_t and ω_c) and the fiber index (β).

$$\frac{h_2}{D} = \left[\frac{1 - 2.4(\omega_t\gamma_t - \omega_c\gamma_c)}{\xi(1 + 2.4\beta)}\right] \quad \text{and} \quad \frac{h_1}{D} = \left(1 - \frac{h_2}{D}\right)$$

The parameter (γ_t and γ_c) in Eq. 7.10 is a constant which describes the extent to which the stress in the conventional longitudinal reinforcement bars, having a yield strength (f_y), is mobilized under the prevailing strain condition in their respective zones of the beam section. Table 7.8 lists different values of the stress-mobilization factor (γ) corresponding to the prevailing strain values in the reinforcing steel bars in the beam section.

Equation 7.10 is depicted graphically in Figs. 7.2, 7.3, 7.4 for the balance beam sections having conventional reinforcement in its tension zone only. Depending upon the position of reinforcing bars in the section, the applicable chart can be selected to determine the bending resistance. The shear capacity of the section can be estimated using Eqs. 7.8 and 7.9. The data given in Table 7.7 can also be used for this purpose. However, it is mandatory to ensure by testing that the concrete containing steel fibers meets the acceptance criterion formulated considering the requirements of fresh concrete properties and the other strength properties.

While using the charts given in Figs. 7.2, 7.3, 7.4 an appropriate combination of reinforcing index (ω) and fiber index (β) should be used to ensure that the section remains balanced or under-reinforced in order to get the highest possible material toughness. For this purpose, the neutral-axis coefficient (h_1/D) given in Eq. 7.10 can be used to find the value of the reinforcing index for any arbitrarily selected value of the fiber index. The balanced beam section has the neutral-axis coefficient

Table 7.8 Values of stress-mobilization factor (γ) for reinforcing rebars having yield strength, f_y = 415 MPa

Strain	<0.00144	0.00144	0.00154	0.00174	0.00214	0.00244	≥0.0038
Factor (γ)	$\varepsilon E/f_y$	0.70	0.74	0.78	0.83	0.85	0.87

Source Author

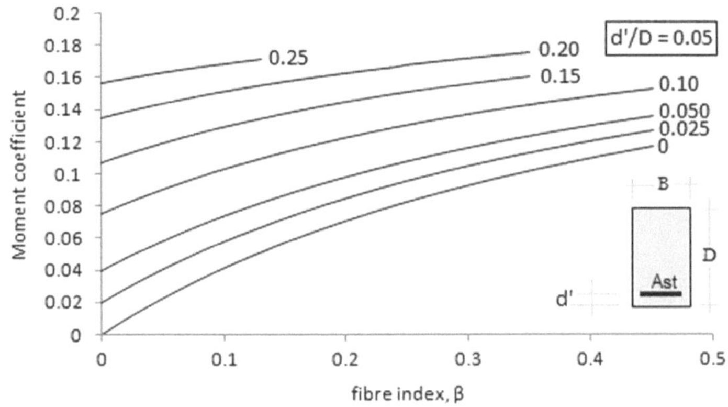

Fig. 7.2 Capacity curve for a concrete beam section, reinforced with conventional tension rebars at a depth ratio of 0.05. *Source* Singh [17]

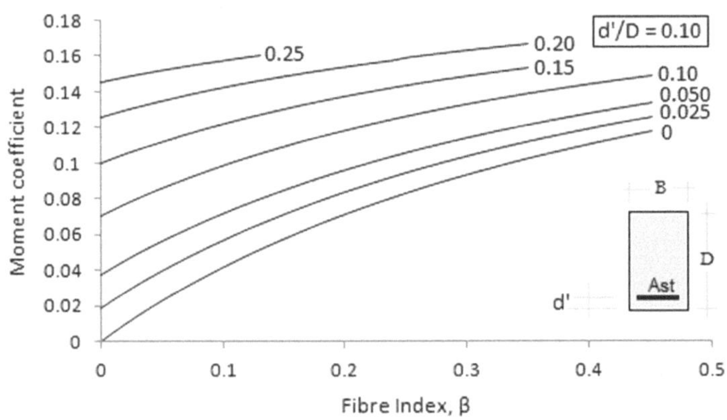

Fig. 7.3 Capacity curve for a concrete beam section, reinforced with conventional tension rebars at a depth ratio of 0.10. *Source* Singh [17]

of 0.53, 0.48 and 0.46 for the Fe250, Fe415 and Fe500 grade steel grades, respectively. These values must be multiplied by the depth ratio (*d*/*D*) before equating with the expression given therein.

Example: A concrete mix (of M25 grade) is being produced by mixing hooked-end fibers at the time of its mixing. The fibers of having aspect ratio 75 are mixed at a rate of 80 kg/m^3. What should be the residual tensile strength of the concrete? And what bending and shear capacity, we will get if a rectangular beam (of size 300 × 450 mm) is cast using this mix.

The dosage of 80 kg/m^3 is corresponding to a volume fraction of 1.02% (= 80/78.5). So, by knowing the *l*/*d* and V_f, the residual tensile strength of SFRC can be obtained from Eq. 7.5a:

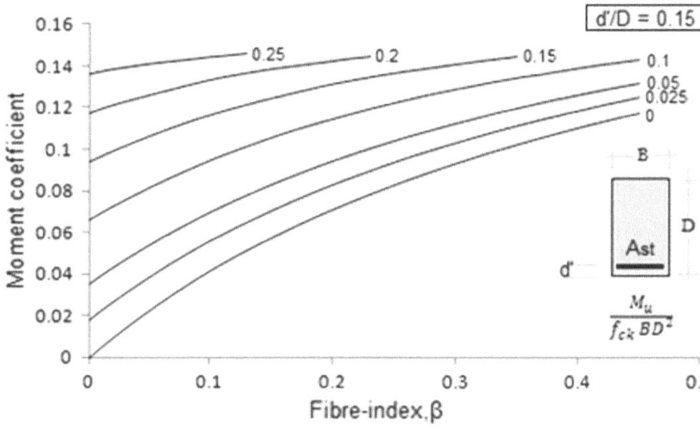

Fig. 7.4 Capacity curve for a concrete beam section, reinforced with conventional tension rebars at a depth ratio of 0.15. *Source* Singh [17]

$$\sigma_t = 0.3 \times 2 \times \sqrt{25} \times \frac{1.02}{100} \times 75 = 2.29 \text{ MPa}$$

Therefore, the fiber index (β) = $\frac{2.29}{25} = 0.0918$.

Assuming the shear span-to-depth (*a/D*) ratio of 3.5, the shear strength of the concrete can be computed by adding Eqs. 7.7 and 7.9, as below:

$$\tau = \frac{kf_c\beta^{0.94}}{3\left(\frac{a}{D}\right)^2} + \frac{(k+1)}{5} f_c\left(\beta - 0.1\alpha\beta^{0.3}\right) = 1.08 \text{ MPa}$$

The value of the constants (k) and (α) in the above calculation is taken as 3.05 and 1, respectively, for the given value of the fiber index (= 0.0918), reinforcement index (= 0) and the concrete compressive strength (= 25 MPa).

The moment capacity of the section can be determined Eq. 7.10, as follows:

$\frac{M_u}{f_{ck}BD^2} = 0.24\left(\frac{h_1}{D}\right)^2 + \{0.5\beta\}\left(\frac{h_2}{D}\right)^2 = 0.0382$, thereby giving a moment capacity of 58.09 kNm for the given beam section of 300 × 450 mm, cast in M25 concrete possessing a residual tensile strength of 2.29 MPa.

The value of the constant (ξ) is taken as unity for SFRC. The neutral-axis coefficients $h_1/D = 0.181$ and $h_2/D = 0.819$ are considered in the above computations; these two values can be determined from Eq. 7.10.

Alternatively, the value of fiber index (***β***) can be easily determined using Eqs. 7.7 to 7.10 to achieve the desired shear and moment section capacity. Subsequently, Eq. 7.5 can be used to find an appropriate combination of the fiber aspect ratio and its volume fraction required to achieve the desired shear and moment capacity.

Example: A rectangular SFRC beam has cross-sectional dimensions of 300 mm × 600 mm. Assuming M40 concrete and Fe 415 grade steel, design a suitable dosage of the hooked-end steel fibers in the concrete required to provide an ultimate moment section capacity (M_u) of 275 kNm alone and then, by considering the contribution of conventional reinforcing bars in its tensile zone. The yield strength of steel fibers is 1150 MPa. What would be the shear capacity of the beam section?

Given: B = 300 mm, D = 600 mm, Take d = 550 mm, f_y = 415 MPa, f_{ck} = 40 MPa, M_u = 275 kNm.

And, the corresponding moment coefficient $= \frac{M_u}{f_{ck}BD^2} = 0.066$.

Equation 7.10 can be used to determine the fiber index (***β***) needed to achieve the required moment capacity; it is found as 0.182 without any conventional longitudinal reinforcement in the section (fiber alone case). The fiber aspect ratio and the corresponding value of volume fraction needed to get a fiber index of 0.182 are obtained from Eq. 7.5 as follows:

$$= 0.3 \times 2 \times \sqrt{40} \times \frac{l}{d} \times V_f = 0.182 \times 40 \quad \text{or} \quad \frac{l}{d} V_f = 1.92$$

The above equation provides a number of options to decide an appropriate combination of the fiber parameters needed to achieve the desired tensile strength; the final value depends upon the achievement of required workability and other characteristics of that mix proportions. Table 7.9 lists all such possible combinations to achieve the desired tensile strength (= 0.182 × 40 MPa). Equations 7.3 and 7.4 are used to decide the critical aspect ratio and the fiber dosage required to activate the fiber pullout failure mode at the ultimate state; these are given below:

$$\left(\frac{l}{d}\right)_c = \frac{0.58 \times 1150}{2\sqrt{40}} = 52.73$$

$$V_f \geq 4\frac{\sqrt{40}}{2 \times 1150} = 1.10\%$$

The fiber aspect ratio therefore should be kept less than its critical value (= 50 say) and any fiber dosage to be used in the mix must be more than 1.10%.

Table 7.9 Design options to select an appropriate combination of fiber parameters

Aspect ratio, l/d	Fiber dosage		Min. volume fraction, V_f (%) needed to ensure strain hardening response (see, Eq. 7.6)
	Volume fraction, V_f (%)	Weight of the steel fibers (kg/m^3)	
40	4.80	377	2.91
50	3.84	301	2.33

Source Author

The tabulated data shows that it is practically impossible to achieve the desired moment capacity by using the fibers alone; there will be considerable workability-related issues if one goes with the high fiber dosage, usually more than 2.5%. Therefore, the only option is to reduce the fiber content by employing conventional steel in the section to share some part of the internal stresses being induced therein.

In all such cases, the SFRC beam section is usually proportioned in its balance state and the remaining value of the moment demand is achieved by providing conventional reinforcing steel bars in the tension zone of the section. The critical fiber index (β_c) for achieving the balanced-state for any SFRC rectangular beam section is 0.1123, and the corresponding moment coefficient is 0.0458 [17]. Therefore, the balanced moment capacity (M_{ub}) of the beam section is as follows:

$$M_{ub} = 0.0458 \times 40 \times 300 \times 600^2 = 197.85 \text{ kNm}$$

The balance flexural capacity (M_{u1}) required to be provided by the conventional longitudinal tensile rebars = 275 − 197.85 = 80 kNm (say).

Therefore, the approximate value of the rebar tensile steel, $A_{st} \cong \frac{M_{u1}}{0.87 f_y d} =$ 402.87 mm^2.

Using 4 No.-12 diameter rebars in the tensile zone of the beam with a following set of the depth ratios meets the area requirement.

- $\frac{d'}{d} = \frac{50}{550} = 0.091$
- $\frac{d_1}{D} = \frac{50}{600} = 0.083$
- $\frac{d}{D} = \frac{550}{600} = 0.917$
- The value of the reinforcement index (ω_t) = $\frac{4\times113\times415}{300\times600\times40} = 0.026$
- For Fe415 steel rebars, the limiting value of the neutral-axis $(h_1/D)_c = 0.48\ (d/D)$ = 0.44
- Assuming an under-reinforced section (i.e., $\gamma = 0.87$)
  $\frac{h_1}{D} = 0.258 < 0.44$ and $\frac{h_2}{D} = 0.742$ (using Eq. 7.10)

The section is therefore under-reinforced. Our assumption of taking $\gamma = 0.87$ was correct so there is no need to iterate and find new value of the neutral-axis coefficients. In some cases, we have to use iterations to reach a true value of the neutral-axis. Knowing this, the strain at the rebar level in the beam section can be determined as follows.

$\varepsilon_{st} = 0.004\left(\frac{D}{h_1}\frac{d}{D} - 1\right)$ = 0.0102 (>0.0038, thereby showing that the beam section is under-reinforced). The corresponding moment of resistance, considering the contribution of both fibers and conventional longitudinal tensile rebars in the section can be determined using Eq. 7.10.

$$\frac{M_u}{f_{ck}BD^2} = \left[0.24\left(\frac{h_1}{D}\right)^2 + \left(\frac{h_2}{D}\right)^2\left\{0.5\beta + \gamma\omega\left(\frac{D}{h_2}\right)\left(1 - \frac{d_1}{D}\frac{D}{h_2}\right)\right\}\right] = 0.0637.$$

$$M_u = 0.0637 \times 40 \times 300 \times 600^2 = 2{,}75{,}184{,}000 \text{ Nmm } (= 275.18 \text{ kNm} > 275 \text{ kNm}).$$

The dosage of the steel fibers required in the beam section can be determined from the fiber index $\boldsymbol{\beta}$ ($= 0.1123$) adopted in the design. Nevertheless, the fiber aspect ratio (***l/d***) finalized in the design should not exceed the critical fiber aspect ratio. The critical fiber aspect ratio $(l/d)_c$ for the present case is computed as 52.73. The maximum fiber aspect ratio, therefore, should be limited to 50 to avoid the fiber fracturing at the ultimate state.

$$\beta = 0.1123 = \frac{0.3V_f(l/d)}{\sqrt{f_{ck}}} = \frac{0.3 \times 2 \times V_f(l/d)}{\sqrt{40}}$$

$$\Rightarrow V_f(l/d) = 1.183.$$

Table 7.10 shows a set of different dosage (V_f) that can be adopted in the concrete mix design to achieve the required moment of resistance along with the 4 No. of 12 mm diameter longitudinal tensile rebars provided at an effective cover of 50 mm from the bottom face of the beam. The fiber length and its diameter can be taken corresponding to the volume fraction (V_f) adopted for the mix.

The computed fiber dosage indicated in Table 7.10 is on a higher side; it may cause mixing and workability issues if used as such. The other options that can be used to avoid any such problems is to try wavy-shaped fibers instead of the hooked-end fibers or to increase the steel percentage (of the longitudinal rebars) in the section; it would result in the reduction of the fiber dosage. Any value of the fiber volume fraction in the range of 0.75–1.75% is okay for most of the construction projects, and it should never be allowed to exceed 2.5%.

For instance, if the value of fiber index is taken as 0.06 (say) and the value of the reinforcing index needed to achieve the balanced section can be computed from Eq. 7.10; it is obtained as 0.072. The value of neutral-axis coefficients for this combination can be obtained as 0.2572 and 0.7427, respectively, for h_1/D and h_2/D. The corresponding moment capacity of the beam section is coming as 317.51 kNm (>275 kNm). In order to obtain this moment capacity, the section

Table 7.10 Design options to select an appropriate combination of fiber parameters

Aspect ratio, *l/d*	Volume fraction, V_f (%)	Steel fibers dosage (kg/m^3)
40	2.96	232
50	2.36	186

Source Author

must be reinforced using longitudinal steel having area of 1249.16 mm^2 (= 0.072 × 40 × 300 × 600/415). Therefore, if 4 no.-20 diameter rebars are provided in the tension zone of the beam, the required reinforcement index can be obtained for the given beam section. The fiber parameters needed in the section are obtained as below:

$$\beta = 0.06 = \frac{0.3V_f(l/d)}{\sqrt{f_{ck}}} = \frac{0.3 \times 2 \times V_f(l/d)}{\sqrt{40}}$$

$$\Rightarrow V_f(l/d) = 0.632$$

The value of the fiber parameters (maybe the aspect ratio or the volume fraction) is given in Table 7.11 for a ready reference.

Table 7.11 indicates that the current fiber dosage will not pose any workability problems during the mixing operations. Still, laboratory investigations are essential to see the response of freshly prepared concrete and deciding upon the fiber parameters that meet the strength as well as constructability issues before fixing any acceptance criterion. It is worth to note that the reduction in the fiber dosage results in the increased requirement of conventional rebars in the section in order to get the same moment capacity. A number of combinations will be at the disposal of the engineers to decide the reinforcement (both in the form of longitudinal steel rebars and fibers) needed in the concrete section that meets the required moment demand; anyone that satisfies the constructability constraints can be chosen.

The shear capacity of the beam section can be determined from Eqs. 7.7 and 7.9. It has been done for a shear span-to-depth (*a/D*) ratio of 3.5, as below:

$$\tau = \frac{kf_c\beta^{0.94}}{3\left(\frac{a}{D}\right)^2} + \frac{(k+1)}{5}f_c\left(\beta - 0.1\alpha\beta^{0.3}\right) + 0.3\sqrt{\rho_t f_c} = 1.95\ \text{MPa}$$

The value of the constants (k) and (α) in the above calculation is taken as 3.69 and 1, respectively, for the given value of the fiber index (= 0.06), α = 0.46 for reinforcement index of 0.072 and the concrete compressive strength (= 40 MPa). The high value of shear strength (= 1.95 MPa) indicates that the beam section without any transverse steel may be sufficient to support the applied load. It is important to highlight that the last term in the above expression shows the contribution from the dowel action of rebars (4 No.-20 diameter) provided in the section.

Table 7.11 Design options to select an appropriate combination of fiber parameters

Aspect ratio, *l/d*	Volume fraction, V_f (%)	Steel fibers dosage (kg/m^3)
40	1.58	124
50	1.26	99

Source Author

One the other hand, if we go for reinforced section using longitudinal steel bars alone, we would need the following steel and Eq. 7.10 can be used to check the moment of resistance of the beam section.
Approximate value of the rebar tensile steel, $A_{st} \cong \frac{275\times10^6}{0.87(0.8)f_y d} = 1731\ \text{mm}^2$.

Using 4 No.-25 diameter rebars in the tensile zone of the beam with a following set of the depth ratios meets the area requirement.

- $\frac{d}{D} = \frac{550}{600} = 0.917$
- $\frac{d_1}{D} = \frac{50}{600} = 0.083$
- The value of the reinforcement index $(\omega_t) = \frac{4\times491\times415}{300\times600\times40} = 0.1132$
- For Fe415 steel rebars, the limiting value of the neutral-axis $(h_1/D)_c = 0.48\ (d/D) = 0.44$
- Assuming an under-reinforced section (i.e., $\gamma = 0.87$)
  $\frac{h_1}{D} = 0.112 < 0.44$ and $\frac{h_2}{D} = 0.888$ (using Eq. 7.10)

The section is therefore under-reinforced. Our assumption of taking $\gamma = 0.87$ was correct so there is no need to iterate and find new value of the neutral-axis coefficients. In some cases, we have to use iterations to reach a true value of the neutral-axis. Knowing this, the strain at the rebar level in the beam section can be determined as follows:

$\varepsilon_{st} = 0.004\left(\frac{D}{h_1}\frac{d}{D} - 1\right) = 0.0287$ (>0.0038, thereby showing that the beam section is under-reinforced). The corresponding moment of resistance, considering the contribution of both fibers and conventional longitudinal tensile rebars in the section can be determined using Eq. 7.10; the fiber index (β) is zero for this case.

$$\frac{M_u}{f_{ck}BD^2} = \left[0.24 \times 0.86\left(\frac{h_1}{D}\right)^2 + \left(\frac{h_2}{D}\right)^2\left\{\gamma\omega\left(\frac{D}{h_2}\right)\left(1 - \frac{d_1}{D}\frac{D}{h_2}\right)\right\}\right] = 0.0819$$

$$\begin{aligned} M_u &= 0.0819 \times 40 \times 300 \times 600^2 \\ &= 353{,}672{,}790\ \text{Nmm}\ (= 353.67\ \text{kNm} > 275\ \text{kNm}). \end{aligned}$$

The shear capacity of the RC beam section can be taken from Table 7.5, corresponding to the percentage steel area of 1.19% provided in the section; it is found as 0.70 MPa. It is provided by the concrete compression zone lying above the neutral-axis along with the dowel action provided by the rebars in the beam tension zone. As this value is small in comparison with the previous case, the section will need transverse steel to make it safe in shear. It is also important to note that RC beams need to be checked for the possible crack-width that would develop on its tensile face unlike the beam reinforced using the steel fibers, where the crack width is usually fine and well scattered over the beam tension face. The RC beams may need tension steel more than needed otherwise to provide the required moment capacity just to keep the crack width within the permitted range, usually less than 0.3 mm.

7.7 Annexure

Some shear strength models available to compute the shear strength of SFRC rectangular beams are tabulated below:

Reference	Details of the prescribed formulae
RELIM [15]	f_{R4} can be adopted from the laboratory results $k_l = 1 + \sqrt{\frac{200}{d}}$ but ≤ 2; $k_f = 1$; ρ_t is percentage longitudinal tension steel in the section $\tau_d = 0.7 k_l k_f [0.12 f_{R4}] + 0.12 k_l (100 \rho_t f_{ck})^{1/3}$
CNR-DT-204 [6]	f_{tuk} (for CMOD of 1.5 mm), $\gamma = 1$ $k = 1 + \sqrt{\frac{200}{d}}$ but ≤ 2; ρ_t is longitudinal steel in the tension zone of beam section $f_{ctk} = 0.7\sqrt{f_c}$ $\tau_u = \frac{0.18k}{\gamma}\left\{\left[100\rho_t\left(1 + 7.5\frac{f_{tuk}}{f_{ctk}}\right)f_{ck}\right]\right\}^{1/3}$
fib MC [1]	$k = 1 + \sqrt{\frac{200}{d}}$ but ≤ 2; ρ_t is longitudinal steel in the tension zone of beam section and $f_{ctk} = 0.7\sqrt{fc}$, $\gamma = 1.5$ $\tau_d = \frac{0.18k}{\gamma}\left\{100p_t f_{ck}\left[\left(1 + 7.5\frac{f_{tuk}}{f_{ctk}}\right)\right]^{\frac{1}{3}}\right\}$
ACI 318 [2]	$\tau_u = \left(0.7\sqrt{f_c'} + 7F\right)\frac{d}{a} + 17.2\rho_t\frac{d}{a}$, F is the product of fiber aspect ratio (l/d) and its volume fraction (V_f)
Arslan [3]	$\tau_d = \left\{0.2 f_c^{2/3}\left(\frac{c}{d}\right) + \sqrt{p_t(1+4F)f_c}\right\}\left(\frac{3}{a/d}\right)^{1/3}$, F is the product of fiber aspect ratio (l/d) and its volume fraction (V_f)
Sharma [16]	$\tau_u = \frac{2}{3}f_t\left(\frac{d}{a}\right)^{0.25}$ where $f_t = 9.5\sqrt{f_c}$ (f_t and f_c in *psi*)
Khuntia et al. [12]	$\tau_u = (0.167e + 0.25F)\sqrt{f_c}$ where e is arch-action factor = 1 for $a/d >$ 2.5, and $e = 2.5d/a$ for $a/d <$ 2.5, F is the product of fiber aspect ratio (l/d) and its volume fraction (V_f)
Narayanan and Darwish [14]	$f_{spfc} = \frac{f_c}{(20 - \sqrt{F})} + 0.7 + \sqrt{F}$ $\tau_{cr} = \left(0.24 f_{spfc} + 0.5F + 20\rho_t\frac{d}{a}\right)$ $\tau_u = e\left(0.24 f_{spfc} + 80\rho_t\frac{d}{a}\right) + 1.7F$ where e is arch-action factor = 1 for $a/d >$ 2.8, and $e = 2.8d/a$ for $a/d <$ 2.8, F is the product of fiber aspect ratio (l/d) and its volume fraction (V_f)
Kwak et al. [13]	$f_{spfc} = \frac{f_c}{(20 - \sqrt{F})} + 0.7 + \sqrt{F}$ and $\tau_{cr} = 3\left(f_{spfc}\right)^{2/3}\left(\frac{\rho_t}{a/d}\right)^{1/3}$, F is the product of fiber aspect ratio (l/d) and its volume fraction (V_f)
Ashour et al. [4]	$\tau_u = \left(2.11\sqrt[3]{f_c'} + 7F\right)\left(\frac{\rho_t}{a/d}\right)^{1/3}$ for $a/d >$ 2.5 and $\tau_u =$ $\left\{\left(2.11\sqrt[3]{f_c'} + 7F\right)\left(\frac{\rho_t}{a/d}\right)^{1/3}\right\}\frac{2.5}{a/d} + 1.7F\left(2.5 - \frac{a}{d}\right)$ for $a/d <$ 2.5, F is the product of fiber aspect ratio (l/d) and its volume fraction (V_f)

References

1. *fib* Model Code for Concrete Structures 2010 (2013) Ernst & Sohn, Berlin, Germany
2. ACI (American Concrete Institute) (2011) ACI 318–11: building code requirements for structural concrete and commentary. ACI, Farmington Hills, MI, USA
3. Arslan G (2014) Shear strength of steel fiber reinforced concrete slender beams. KSCEJ Civ Eng 18:587
4. Ashour SA, Hasanain GS, Wafa FF (1992) Shear behaviour of high strength fiber reinforced concrete beams. ACI Struct J 89(2):176–184
5. BIS (2000) IS 456: code of practice for the design of reinforced concrete structures. BIS, New Delhi, India
6. CNR DT 204 (2006) Guide for the design and construction of fiber-reinforced concrete structures. National Research Council, Rome
7. Hendry AW (1990) Structural masonry. MacMillan Education Ltd., London
8. IS 1121 (1975) Indian Standard Code of practice for method of test for determination of strength properties of natural building stones: part 1 (compressive strength)
9. IS 1905 (1987) Indian Standard Code of practice for structural use of unreinforced masonry
10. IS 2250 (1981) Code of practice for preparation and use of masonry mortars
11. IS 3629 (1986) Specification for structural timber in building (first revision)
12. Khuntia M, Stojadinovic B, Goel S (1999) Shear strength of normal and high-strength fiber reinforced concrete beams without stirrups. ACI Struct J 96(2):282–290
13. Kwak Y, Eberhard MO, Kim W, Kim J (2002) Shear strength of steel fiber-reinforced concrete beams without stirrups. ACI Struct J 99(4):530–538
14. Narayanan R, Darwish IYS (1987) Use of steel fibers as shear reinforcement. ACI Struct J 84 (3):216–227
15. RILEM TC 162-TDF (2003) Test and design methods for steel fibre reinforced concrete σ-ε-design method. Mater Struct 36:560–567
16. Sharma AK (1986) Shear strength of steel fiber reinforced concrete beams. ACI Struct J 83 (4):624–628
17. Singh H (2017) Steel fiber reinforced concrete: behavior, modelling and design. Springer Transactions in Civil and Environmental Engineering, Singapore
18. Singh H (2020) Closed-form solution for shear strength of steel fiber-reinforced concrete beams. ACI Struct J 117(3):261–272.

Index

H. Singh, *Structural Materials*, Materials Horizons: From Nature to Nanomaterials,
https://doi.org/10.1007/978-981-16-3211-2